Lecture Notes in Control and Information Sciences 424

Editors: M. Thoma, F. Allgöwer, M. Morari

Maria Letizia Corradini, Andrea Cristofaro,
Fabio Giannoni, and Giuseppe Orlando

Control Systems with Saturating Inputs

Analysis Tools and Advanced Design

Springer

Authors
Prof. Maria Letizia Corradini
Università di Camerino
Scuola di Scienze e Tecnologie
Via Madonna delle Carceri
Camerino
Italy

Prof. Fabio Giannoni
Università di Camerino
Scuola di Scienze e Tecnologie
Via Madonna delle Carceri
Camerino
Italy

Dr. Andrea Cristofaro
Università di Camerino
Scuola di Scienze e Tecnologie
Via Madonna delle Carceri
Camerino
Italy

Dr. Giuseppe Orlando
Dipartimento di ingegneria
dell'informazione
Università Politecnica delle Marche
Ancona
Italy

ISSN 0170-8643 e-ISSN 1610-7411
ISBN 978-1-4471-2505-1 e-ISBN 978-1-4471-2506-8
DOI 10.1007/978-1-4471-2506-8
Springer London Heidelberg New York Dordrecht

Library of Congress Control Number: 2011945434

Printed on acid-free paper

Springer is part of Springer Science+Business Media (www.springer.com)

Preface

Saturation nonlinearities are ubiquitous in engineering systems: every physical actuator or sensor is subject to saturation owing to its maximum and minimum limits. Input saturation is an operating condition that is well known to the control community for its side effects, which cause conventional controllers to loose their closed loop performance as well as control authority in stabilization. Therefore, practical application of control theory cannot avoid taking into account saturation nonlinearities in actuators, explicitly dealing with constraints in control design.

This book investigates the problem of actuator saturation from a practice-oriented alternative viewpoint. Analysis tools applicable to plants of arbitrary finite dimension are given providing an analytical estimate of the maximal null controllable region. Nonlinear control design techniques are presented with particular reference to robustness with respect to matched disturbances and/or uncertainties. Design approaches explicitly developed in the discrete-time framework are described in order to enhance the practical applicability of controllers. More specifically, the results that are to be presented in this book are outlined as follows.

In Chapter 1, after a short introduction to linear systems with saturation nonlinearities in the actuator, we state and discuss some definitions and technical terms which will be useful in the following.

Chapter 2 and 3 give explicit descriptions of the null controllable regions of a linear system driven by saturating actuators in the continuous-time and discrete-time framework respectively. The description of the null controllable region is obtained following an iterative procedure based on reversed-time evolution and convexification. In both cases, Single Input planar case are addressed first, then results are extended to n-dimensional Single Input plants and finally to Multi Input Systems.

Chapter 4 addresses design issues in the continuous-time framework. First a design technique is proposed for linear plants subject to saturating actuators, such that the resulting linear controller has the property of having non-increasing norm along the closed-loop system trajectories. In particular the region of attraction associated to the saturating control is an unbounded strip and it can be straightforwardly characterized. Moreover it is shown how, once the saturation level is fixed, it is possible to split the controller into a finite number of saturating components. The number of components can be a priori determined for any fixed compact set of initial data. Next, the problem of controlling uncertain Multi-Input linear plant with saturating actuators is looked at from a different perspective. The objective is to construct time-varying feedback laws, derived imposing the achievement of a sliding motion onto a suitable time-varying sliding surface, able to ensure that saturation thresholds are never violated. It is proved there that a constructive procedure exists for designing the surface as to guarantee the asymptotical stabilization of the plant in the presence of bounded matched uncertainties, under the usual assumption of the saturation threshold being larger than the bound on uncertainties.

The discrete-time counterparts of the approaches described in Chapter 4 are addressed in Chapter 5. It worths noticing that completely different technical solutions are required with respect to the continuous time case. The main features of the proposed sliding-mode based control law are: *i*) no restrictions are needed in the plant structure; *ii*) bounded matched disturbances are considered; *iii*) robust practical stabilization on the null controllability region can be achieved by means of a time-varying state feedback controller, derived imposing the achievement of a quasi-sliding motion onto a suitable time-varying quasi-sliding surface; *iv*) performing transient shaping is not subject to any condition and can be achieved simply by manipulating the dynamics imposed onto the quasi-sliding surface. It is proved that a constructive systematic procedure exists for designing the surface as to guarantee the ultimately boundedness of plant trajectories in the presence of bounded matched uncertainties.

Wherever possible, the various topics are rounded out with results obtained through experiments with actual plants. Mathematical details which are outside the normal province of control engineers are presented in an appendix for the interested reader. The ideas formulated in this book could be of great practical help to professionals, researchers, practitioners and graduate students in control, electrical, and mechanical engineering, working directly with problems related to the control of plants with saturating actuators. Some first–year graduate courses in linear systems and multivariable control or some background in nonlinear control systems would greatly facilitate the reading of this book.

Camerino, Italy
November, 2011

Maria Letizia Corradini
Andrea Cristofaro
Fabio Giannoni
Giuseppe Orlando

Symbols and Notation

$\mathbb{R}$ the field of real numbers

$||\cdot||$ the Euclidean norm

$\mathbf{A}$ a matrix defined in the $n \times m$-dimensional real vector space, $n, m > 1$

$\mathbf{b}$ a vector defined in the r-dimensional real vector space

$\mathbf{x}$ the state vector

x_i the i-th element of the state vector

sat the saturation function for Single Input systems

u the generic unconstrained control input

v the generic constrained control input

sat the saturation function for Multi Input systems

λ a generic eigenvalue

$\mathscr{B}_M$ the largest null controllable region for Single Input systems (continuous-time case)

$\mathscr{B}_{\mathbf{M}}$ the largest null controllable region for Multi Input systems (continuous-time case)

$\mathscr{B}_M^\sharp$ the largest null controllable region for Single Input systems (discrete-time case)

$\mathscr{U}_M$ the set of admissible control inputs (continuous-time case)

$\mathscr{U}_M^*$ the set of bang-bang control inputs (continuous-time case)

$\mathscr{W}_M$ the set of admissible control inputs (discrete-time case)

$\mathscr{W}_M^*$ the set of bang-bang control inputs (discrete-time case)

C^0 the class of continuous functions

$C^0_\sharp$ the class of piecewise continuous functions

ℓ^∞ the set of bounded sequences

$\ell^\infty(\mathbb{Z})$ the set of bounded bilateral sequences

j the imaginary unit

Acronyms

ANCBC	Asymptotically Null Controllable with Bounded Controls
ARE	Algebric Riccati Equation
LHP	Left Half Plane
MI	Multi Input
MIMO	Multi Input, Multi Output
PWC	Piecewise Constant
REM	Maximum Region of Reachability
RHP	Right Half Plane
RRM	Maximum Region of Recoverability
SI	Single Input
SISO	Single Input, Single Output
SM	Sliding Mode
VSS	Variable Structure Systems

Contents

Part I
Introduction and Analysis Tools

Chapter 1
Introduction

1.1 Linear Plants with Actuator Saturation

The presence of actuator saturation in control systems, though frequently ignored, is due to inherent (and unavoidable) physical limitations of devices. The relevance of this issue from the practical viewpoint is more and more attracting the attention of control system researchers, as failure in accounting for actuator saturation may lead to severe deterioration of closed loop system performance, even to instability.

As a matter of fact, classical control theory suffers from the limitation that, in reality, dynamical systems are frequently affected by a number of ignored factors which may invalidate, or at least severely limit its effective application. Among these factors, inherent saturation effects turn out to be almost ubiquitous, since the output of all physical devices are limited to some degree, and when a limiting value is reached saturation is said to occur. A classical example is the output torque of a servomotor, which is subject to a maximum value which can be supplied.

In general, the performance of a closed-loop system where control design has been carried out ignoring saturation may seriously deteriorate. A well known example is the case of the PID controller applied in closed loop in the presence of saturation, producing the so called integrator wind-up. What happens is that, whenever the actuator saturates, the error variable is continuously integrated and this produces larger and larger values of the control input, exceeding the saturation threshold. The schemes known as anti-windup approaches are aimed at mitigating the adverse effects of saturation by usually adding extra feedback loops in order to constrain the control variable to vary within an assigned range.

There are several possible models for saturating devices: a simple model given by the truncation operator is here considered:

$$T_k(s) = \max\{-k, \min\{k, s\}\}, \tag{1.1}$$

where $k \in \mathbb{R}$ is the saturation level. The function $T_k(s)$ is symmetric, linear on a suitable interval containing the origin and constant outside this interval (see Fig. 1.1). Linear systems subject to asymmetric input saturation have also been

M.L. Corradini et al.: Control Systems with Saturating Inputs, LNCIS 424, pp. 3–6.
springerlink.com

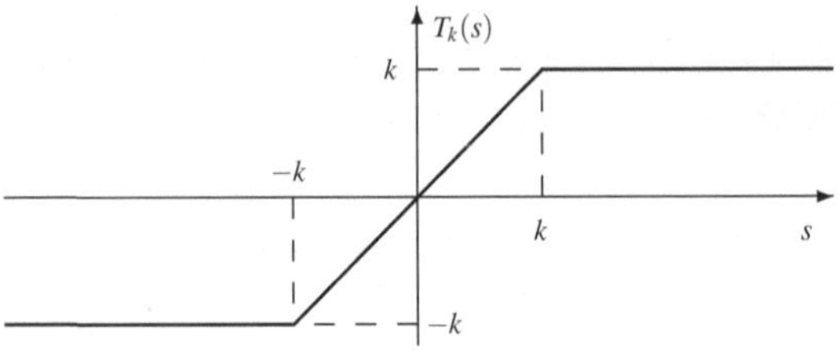

Fig. 1.1. Saturation function.

studied (see e.g. [1]). Denoting with $\mathbf{A} \in \mathbb{R}^{n\times n}$ the state matrix, with $\mathbf{B} \in \mathbb{R}^{n\times 1}$ the input matrix, with $\mathbf{x} \in \mathbb{R}^n$ the state variable and with $u \in \mathbb{R}$ the control input, a SISO linear system subject to actuator saturation can be written as

$$\dot{\mathbf{x}} = \mathbf{A}\mathbf{x} + \mathbf{B}u, \quad \mathbf{x}(0) = \mathbf{x}_0, \tag{1.2}$$

with $u = T_M(v)$, where the constant $M > 0$ is the saturation level.

Throughout this book, the saturation function for a scalar $s \in \mathbb{R}$ will be denoted as

$$\mathrm{sat}_M(s) = T_M(s) \tag{1.3}$$

with a saturation level M, unless differently specified. With a slight abuse of notation, for a vector $\mathbf{s} = [s_1\ s_2 \ldots s_m]^T \in \mathbb{R}^m$, the saturation function will be denoted as

$$\mathbf{sat}(\mathbf{s}) = [\mathrm{sat}(s_1)\ \mathrm{sat}(s_2) \ldots \mathrm{sat}(s_m)]^T \tag{1.4}$$

1.2 Background

Following [2], a system is said to be globally Asymptotically Null Controllable by Bounded Controls (ANCBC) if, for a given bound on the controls, every state can be driven to the origin either in a finite time or asymptotically by a bounded control. Since it has been proved [3] that a linear stabilizable system having all its poles in the closed left-half plane is asymptotically null controllable, such system is said ANCBC.

Existing results pertaining to the problem of global and semi-global stabilization showed that, in general, global stabilization of a linear system subject to saturating actuators can be achieved by a nonlinear controller if and only if the linear system (without saturation) is asymptotically null controllable by bounded control, this condition being equivalent to classical stabilizability and to the requirement that the plant has stable open loop poles [4] [3] [5] [6] [7]. In this framework, a nested feedback technique for designing nonlinear globally asymptotically stabilizing controllers was first proposed for a chain of integrators [8] and then fully generalized [9].

In the semi-global framework, early works proved that, under the appropriate conditions, stabilization of asymptotically null controllable with bounded control linear systems can be achieved using linear feedback laws [10] [11]. To this purpose, a number of design methods for low gain controllers have been proposed, as well as the so called low-and-high design technique [12]. An LQR/LQG theory for systems with saturating actuators has been recently presented [13]. An extensive bibliography on the subject is reported in [14].

If the assumption of asymptotically null controllability with bounded control is relaxed, however, results on stabilizability of uncertain linear systems are more fragmentary. Studying the null controllable region, [15] recently showed that, if a planar antistable plant is considered, for any given set in the null controllable region there exists a saturated linear feedback controller yielding a closed loop system having an asymptotically stable equilibrium whose domain of attraction includes this set.

Linear plants with bounded controls have been widely studied in the field of anti-windup approach [16] [17], where an additional feedback loop is introduced. In most cases, this technique requires both plant stability and/or complete knowledge even for linear time-invariant systems. Exponentially unstable systems have been also considered by [18] in the presence of reference and disturbance inputs. The disturbances, however, are restricted not to act on the exponentially unstable modes. Composite nonlinear feedback control has been recently used [19] for generic linear systems without uncertainties.

In the wide literature addressing the problem of disturbance rejection for linear systems subject to actuator saturation, an interesting research line considers disturbances that are magnitude bounded. This line complements the research thrust addressing $\mathscr{L}_p$ disturbances [20] [21]. In the former framework, [22] proved that semiglobal practical stabilization for a linear system subject to actuator saturation and input additive disturbances can be achieved as long as the open loop system is not exponentially unstable. For the same class of systems, Lin [7] constructed nonlinear feedback laws that achieve global practical stabilization. Very recently, it has been proved in [23] that a 2-dimensional linear systems subject to actuator saturation and bounded input additive disturbances can be globally practically stabilized by linear state feedback.

1.3 Outline of the Book

This book investigates the problem of actuator saturation from a practice-oriented viewpoint. Analysis tools applicable to plants of arbitrary finite dimension are given for the determination of an analytical estimate of the maximal null controllable region. Nonlinear control design techniques are presented with particular reference to robustness with respect to matched disturbances and/or uncertainties. Design approaches explicitly developed in the discrete-time framework are described in order to enhance the practical applicability of controllers. More specifically, the results that will be reported in this book can be outlined as follows.

After having presented, in the present Chapter, a short introduction to linear systems with saturation nonlinearities in the actuator, Chapter 2 and 3 provide the explicit description of the null controllable region of a linear system driven by saturating actuators in the continuous-time and discrete-time framework respectively. In both cases, Single Input planar systems are addressed first, then results are extended to n-dimensional Single Input plants and finally to Multi Input systems. It is worth noting that characterizations of the null controllable regions have been already proposed in the literature in recent years (e.g. in terms of trajectories corresponding to bang-bang controls with a finite number of switchings [24] for a continuous-time system, in the framework of discrete-time systems [25], rational systems [26] and MPC controllers [27]). Since the problem requires a high computational burden, an alternative characterization is proposed in these Chapters, where the description of the null controllable region is made following an iterative procedure based on reversed-time evolution and convexification.

Chapter 4 addresses design issues in the continuous-time framework. First a design technique is proposed for linear plants subject to saturating actuators, such that the resulting linear controller has the property of having non-increasing norm along the closed-loop system trajectories. In particular the region of attraction associated to the saturating control is an unbounded strip and it can be straightforwardly characterized. Moreover it is shown how, once the saturation level is fixed, it is possible to split the controller into a finite number of saturating components. The number of components can be a priori determined for any fixed compact set of initial data. Next, the problem of controlling uncertain Multi-Input linear plant with saturating actuators is looked at from a different perspective. The objective is to construct time-varying feedback laws, derived imposing the achievement of a sliding motion onto a suitable time-varying sliding surface, able to ensure that saturation thresholds are never violated. It is proved there that a constructive procedure exists for designing the surface as to guarantee the asymptotical stabilization of the plant in the presence of bounded matched uncertainties, under the usual assumption of the saturation threshold being larger than the bound on uncertainties. Finally, the discrete-time counterparts of the approaches described in Chapter 4 are addressed in Chapter 5.

Chapter 2
Estimation of the Null Controllable Region: Continuous-Time Plants

2.1 Introductory Remarks and Definitions

The description of the maximal stability region for a constrained system is a fundamental problem in control theory and a huge amount of literature pertaining the subject can be found (see for instance [5], [7], [28], [29], [30], [24]). A seminal work in this field is represented by the paper [31], which introduces *recoverable* and *reachable zones* for linear systems subject to input constraints and exploits the main properties of such regions.

A general constrained control problem can be formulated imposing that the input variable u is allowed to vary only in a prescribed, possibly bounded, subset $\mathscr{U}$, the set of admissible controls. Let us consider the linear continuous-time plant

$$\begin{cases} \dot{\mathbf{x}} = \mathbf{A}\mathbf{x} + \mathbf{B}u \\ \mathbf{x}(0) = \mathbf{x}_0 \end{cases} \tag{2.1}$$

and let us denote by $\phi(t, \mathbf{x}_0, u)$ the solution of the system for $t \in \mathbb{R}$. A saturating device is assumed to be acting on the system, i.e. the input u is subject to the constraint

$$u = \mathrm{sat}_M(v), \; M > 0 \tag{2.2}$$

with

$$\mathrm{sat}_h(s) = \max\{-h, \min\{h, s\}\} \tag{2.3}$$

where the input u is available for direct manipulation. An element $\mathbf{x}_0$ in the state space $\mathbb{R}^n$ is a *recoverable state* in the time interval (t_0, t_f) with respect to the final state $\mathbf{x}_f$, if there exists an admissible control which will drive the system from $\mathbf{x}_0$ at time t_0 to state $\mathbf{x}_f$ at time t_f. The *maximum region of recoverability* with respect to $\mathbf{x}_f$ in (t_0, t_f), denoted by $\mathrm{RRM}(\mathbf{x}_f; t_0, t_f)$, is the set of all recoverable states in (t_0, t_f) with respect to $\mathbf{x}_f$.

An element $\mathbf{x}_f$ in the state space $\mathbb{R}^n$ is a *reachable state* in the time interval (t_0, t_f) with respect to the initial state $\mathbf{x}_0$, if there exists an admissible control which

M.L. Corradini et al.: Control Systems with Saturating Inputs, LNCIS 424, pp. 7–32.
springerlink.com

will drive the system from $\mathbf{x}_0$ at time t_0 to $\mathbf{x}_f$ at time t_f. The *maximum region of reachability* with respect to $\mathbf{x}_0$ in (t_0,t_f), denoted by $\mathrm{REM}(\mathbf{x}_0;t_0,t_f)$, is the set of all reachable states in (t_0,t_f) with respect to $\mathbf{x}_0$.

The regions can be expressed as follows

$$\mathrm{RRM}(\mathbf{x}_f;t_0,t_f) = \left\{ \mathbf{x} \in \mathbb{R}^n : \mathbf{x} = \mathrm{e}^{\mathbf{A}(t_f-t_0)}\mathbf{x}_f - \int_{t_0}^{t_f} \mathrm{e}^{\mathbf{A}(\xi-t_0)}\mathbf{B}u(\xi)d\xi, u \in \mathscr{U} \right\}$$

$$\mathrm{REM}(\mathbf{x}_0;t_0,t_f) = \left\{ \mathbf{x} \in \mathbb{R}^n : \mathbf{x} = \mathrm{e}^{\mathbf{A}(t_0-t_f)}\mathbf{x}_0 + \int_{t_0}^{t_f} \mathrm{e}^{\mathbf{A}(\xi-t_f)}\mathbf{B}u(\xi)d\xi, u \in \mathscr{U} \right\}.$$

If $t_0 = 0$ and $\mathbf{x}_f = 0$, or $\mathbf{x}_0 = 0$, we use the notation of $\mathrm{RRM}(t_f)$ and $\mathrm{REM}(t_f)$, respectively.

Denoting by REM^- the reachable set for the reverse time system, the following result holds.

Proposition 2.1. *The maximum recoverable set of the positive time system coincides with the maximum reachable set of the negative time system*

$$\mathrm{RRM}^+(t_f) = \mathrm{REM}^-(t_f).$$

Thanks to the Jordan canonical decomposition, the system can be always split into a stable and an unstable subsystem. Suppose that the state matrix $\mathbf{A}$ has m eigenvalues with positive real part and $n-m$ eigenvalues with non-positive real part; then, up to a coordinates change, the system can be written in the equivalent form

$$\begin{pmatrix} \dot{\mathbf{x}}_u \\ \dot{\mathbf{x}}_s \end{pmatrix} = \begin{pmatrix} \mathbf{A}_u & 0 \\ 0 & \mathbf{A}_s \end{pmatrix} \begin{pmatrix} \mathbf{x}_u \\ \mathbf{x}_s \end{pmatrix} + \begin{pmatrix} \mathbf{B}_u \\ \mathbf{B}_s \end{pmatrix} u, \tag{2.4}$$

where $\mathbf{x}_u, \mathbf{B}_u \in \mathbb{R}^m$, $\mathbf{x}_s, \mathbf{B}_s \in \mathbb{R}^{n-m}$, $\mathbf{A}_u \in \mathbb{R}^{m\times m}$ and $\mathbf{A}_s \in \mathbb{R}^{n-m\times n-m}$.

The following theorem on the maximum recoverable region has been proved by J. LeMay in [31].

Theorem 2.1. *Given the system (2.4), the asymptotic maximum recoverable region ($t_f = +\infty$) is given by*

$$\mathrm{RRM}(\infty) = \mathrm{RRM}_u(\infty) \times \mathbb{R}^{n-m},$$

where $\mathrm{RRM}_u(\infty)$ *is the asymptotic maximum recoverable region of the system*

$$\dot{\mathbf{x}}_u = \mathbf{A}_u\mathbf{x}_u + \mathbf{B}_u u.$$

The theorem states that the recoverable region is completely determined by the unstable subsystem; in particular if the system has m unstable modes, the recoverable region contains a linear subspace of dimension $n-m$.

In the presence of saturation constraints, the recoverable region is often denominated *null controllable region*, which is the largest set of initial data ensuring asymptotic stability for a saturating system with an admissible control function. In order to define it rigorously we give some definitions.

Definition 1. Given a saturation level $M > 0$, the set of admissible control inputs is defined as

$$\mathscr{U}_M = \left\{ u \in \mathrm{C}^0_\sharp(\mathbb{R}) : \|u\|_\infty = \sup_{t\in\mathbb{R}} |u(t)| \leq M \right\}.$$

Definition 2. Given the system (2.1) we define *saturated maximal region of attraction* the set

$$\mathscr{B}_M = \left\{ x \in \mathbb{R}^n : \exists u \in \mathscr{U}_M \text{ with } \lim_{t\to\infty} \|\varphi(t,x,u)\| = 0 \right\}.$$

Definition 3. The set $\mathscr{U}_M^* \subset \mathscr{U}_M$ is the set of admissible bang-bang control functions

$$\mathscr{U}_M^* = \{u \in \mathscr{U}_M : u(t) \in \{M, -M\} \ \forall t \in \mathbb{R}\}.$$

Defining $\mathscr{B}_u$ as the region of attraction associated to the control variable u, the following identity holds

$$\mathscr{B}_M = \bigcup_{u\in\mathscr{U}_M} \mathscr{B}_u.$$

The problem of stabilization of linear systems with bounded controls has been extensively studied and completely solved for ANCBC systems (see [5], [6], [7]). In particular if all the eigenvalues of the state matrix have non positive real part, the saturated maximal region of attraction is the whole space; moreover semi-global stabilization for such systems can be achieved by a linear feedback.

For systems having exponentially unstable modes, thanks to Theorem 2.1, it is enough to study the null controllable region of the antistable subsystem. In [24], the authors present a detailed analysis of the null controllable regions based on control inputs having a finite number of switching from maximum and minimum admissible values. The approach described in the present chapter is different (see [32]): starting from the knowledge of controlled invariant strips for the saturated system, thanks to an iterated convexification technique, a recursive sequence of sets is defined and the null controllable region is obtained as limit of such approximating sequence. The first part of the chapter is dedicated to single input systems, while in the last section the MIMO case is presented.

2.2 The SISO Planar Case

Consider a 2-dimensional controllable SISO linear system having, without loss of generality, the following structure

$$\begin{cases} \dot{\mathbf{x}} = \mathbf{A}\mathbf{x} + \mathbf{B}u \\ \mathbf{x}(0) = \mathbf{x}_0 \end{cases} \tag{2.5}$$

where

$$\mathbf{A} = \begin{pmatrix} 0 & 1 \\ a_1 & a_2 \end{pmatrix}, \qquad \mathbf{B} = \begin{pmatrix} 0 \\ 1 \end{pmatrix},$$

$\mathbf{x}(t) = (x_1(t), x_2(t)) \in \mathrm{C}^0(\mathbb{R}, \mathbb{R}^2)$ is the state vector and $v(t) \in \mathrm{C}^0_\sharp(\mathbb{R})$ is the input variable. Input saturation is assumed for the system; in particular the input v is subject to the constraint (2.2).

2.2.1 Real Eigenvalues

For a planar system having real unstable eigenvalues it is possible to determine an invariant (controlled) parallelogram as external bound to the null controllable region. Moreover such bound is sharp, in the sense that the null controllable region and the parallelogram are tangent one to the other; setting the control u equal to the saturated values $\pm M$ and considering reversed-time evolution with the corner points of the parallelogram as initial positions, the boundary of $\mathscr{B}_M$ is completely covered by the corresponding system trajectories.

Consider the system (2.5) in the case when the state matrix $\mathbf{A}$ has real eigenvalues $\lambda_1, \lambda_2 \in \mathbb{R}$. Let us define the functions

$$\psi_{\pm M}(t) = x_1(t) + m x_2(t) + q; \tag{2.6}$$

In order to determine constants $m, q \in \mathbb{R}$ such that, when the control variable u is taken identically equal to $\pm M$, we have

$$\psi_{\pm M}(t) = 0 \Rightarrow \dot{\psi}_{\pm M}(t) = 0.$$

Since

$$\begin{aligned} \dot{\psi}_{\pm M}(t) &= \dot{x}_1(t) + m\dot{x}_2(t) \\ &= x_2(t) + m(a_1 x_1(t) + a_2 x_2(t) \pm M); \end{aligned} \tag{2.7}$$

imposing $\psi_{\pm M}(t) = 0$ we get $x_1(t) = -m x_2(t) - q$. The equation $\dot{\psi}_{\pm M}(t) = 0$ can be written as

$$\dot{\psi}_{\pm M}(t) = x_2(t) + m(-a_1 m x_2(t) - a_1 q + a_2 x_2(t) \pm M) = 0.$$

and we obtain

$$\begin{cases} q = \pm \dfrac{M}{a_1} \\ \\ a_1 m^2 - a_2 m - 1 = 0. \end{cases}$$

From the last equation we get

$$m = \frac{a_2 \pm \sqrt{a_2^2 + 4a_1}}{2a_1}; \tag{2.8}$$

the expression $a_2^2+4a_1$ is non negative if and only if A has real eigenvalues. In particular we have

$$m=\frac{\lambda_1+\lambda_2\pm|\lambda_1-\lambda_2|}{-2\lambda_1\lambda_2}\in\left\{-\frac{1}{\lambda_1},-\frac{1}{\lambda_2}\right\}.$$

We denote by E_M^1, E_M^2 the (unbounded) strips

$$E_M^1=\left\{\mathbf{x}\in\mathbb{R}^2:\left|x_1-\frac{1}{\lambda_1}x_2\right|<\frac{M}{|\lambda_1\lambda_2|}\right\}$$

$$E_M^2=\left\{\mathbf{x}\in\mathbb{R}^2:\left|x_1-\frac{1}{\lambda_2}x_2\right|<\frac{M}{|\lambda_1\lambda_2|}\right\}$$

Let us denote by $\Gamma_{\pm M}$ the trajectory of $\varphi(t,\pm M/a_1,\pm M)$ for $t\in(-\infty,0]$.

Remark 2.1. *Note that the two curves Γ_M and Γ_{-M} are symmetric with respect to the origin by construction.*

Theorem 2.2. *Given the system (2.5) with $\lambda_1,\lambda_2\in\mathbb{R}$, $\lambda_1>0$ and $\lambda_2>0$, the set $\Gamma_M\cup\Gamma_{-M}$ can be regarded as the boundary of a domain D. Moreover D coincides with the saturated maximal region of attraction $\mathscr{B}_M$ and we have*

$$\partial\mathscr{B}_M=\Gamma_M\cup\Gamma_{-M}.$$

Proof. We give the proof for $\lambda_1\neq\lambda_2$. Since $a_1=-\lambda_1\lambda_2<0$ and $\lambda_1,\lambda_2>0$, from (2.7) we obtain for $j=1,2$

$$\psi_{\pm M}^{\lambda_j}(t)<0\Rightarrow\dot{\psi}_{\pm M}^{\lambda_j}(t)<0$$

$$\psi_{\pm M}^{\lambda_j}(t)>0\Rightarrow\dot{\psi}_{\pm M}^{\lambda_j}(t)>0,$$

where $\psi_{\pm M}^{\lambda_j}(t)$ is the function defined by (2.6)-(2.7) and associated to the coefficient $m=1/\lambda_j$, $j=1,2$; as a consequence we have

$$\mathscr{B}_M\subseteq E_M^1\cap E_M^2.$$

Moreover, since for $x_2<0$ we have $\dot{x}_1<0$ and for $x_2>0$ we have $\dot{x}_1>0$, the set of points

$$\left\{\mathbf{x}\in E_1\cap E_2:|x_1|\geq\frac{M}{|\lambda_1\lambda_2|}\right\}$$

is not included in $\mathscr{B}_M$. By contradiction, let us suppose that $(x_1,x_2)=(-M/|a_1|,s)\in\mathscr{B}_M$ with $s<0$. In order to reach the origin, the derivative of the first coordinate of the solution must change sign from negative to positive; this implies that the trajectory must have an intersection with the x_1-axis in $\bar{\mathbf{x}} = (x_1, 0)$, with $x_1 < -M/|a_1|$, that is $\bar{\mathbf{x}} \notin E_1 \cap E_2$. We can conclude that

$$\mathscr{B}_M \subseteq \{\mathbf{x} \in E_1 \cap E_2 : |x_1| < M/|a_1|\}.$$

For $u \equiv \pm M$ the general structure of the system solution $\varphi(t, \mathbf{x}_0, \pm M)$ is given by

$$x_1(t) = C_1(\mathbf{x}_0, M)\exp(\lambda_1 t) + C_2(\mathbf{x}_0, M)\exp(\lambda_2 t) \mp \frac{M}{a_1}$$

$$x_2(t) = \lambda_1 C_1(\mathbf{x}_0, M)\exp(\lambda_1 t) + \lambda_2 C_2(\mathbf{x}_0, M)\exp(\lambda_2 t),$$

where C_1, C_2 are constants depending only on $\mathbf{x}_0$ and M; we see that, for any choice of the initial datum $\mathbf{x}_0$, the solution verifies

$$\lim_{t \to -\infty} \varphi(t, x_0, \pm M) = \left(\mp \frac{M}{a_1}, 0\right).$$

Let us fix $\mathbf{x}_0 = (M/a_1, 0)$ and consider the solution $\varphi(t, \mathbf{x}_0, M)$. Recall that, as a consequence of the unicity of the solution, there is no intersection between the integral curves. Let us denote by Γ_M the trajectory of $\varphi(t, \mathbf{x}_0, M)$ for $t \in [-\infty, 0]$. Γ_M is a regular curve contained in the half-space $\{x_2 \leq 0\}$ which separate into two disconnected components the region $\{\mathbf{x} \in E_1 \cap E_2 : x_2 \leq 0\}$. Let us choose $\mathbf{z}_0$ belonging to the lower part of $\{\mathbf{x} \in E_1 \cap E_2 : x_2 \leq 0\}$; the solution starting from $\mathbf{z}_0$ converges to $(-M/a_1, 0)$ for $t \to -\infty$ by construction and since the trajectory must have no intersection with Γ_M, there exists $T > 0$ such that $\varphi(T, \mathbf{z}_0, M) \notin E_1 \cap E_2$. On the other hand, if $\mathbf{z}_0$ is taken in the region between Γ_M and the x_1-axis, the trajectory of $\varphi(t, \mathbf{z}_0, M)$ intersects the x_1-axis in some $\bar{x} = (\bar{x}_1, 0)$ with $|x_1| < M/|a_1|$. It is easy to see that, as θ varies in $(0, M)$, the curves Γ_θ describe the whole region between Γ_M and the x_1-axis. We can conclude that the boundary of $\mathscr{B}_M$ is $\Gamma_M \cup \Gamma_{-M}$, where Γ_{-M} is the trajectory of the solution $\varphi(t, \mathbf{x}_0, -M)$ for $t \in [-\infty, 0]$ with $\mathbf{x}_0 = (-M/a_1, 0)$. □

Remark 2.2. *The curves Γ_M and Γ_{-M} constitute the boundary of the saturated maximal region of attraction also in the special case of coinciding real eigenvalues $\lambda_1 = \lambda_2 = \lambda > 0$. Note that in this case we have only one invariant strip E_M^λ defined as*

$$E_M^\lambda = \left\{\mathbf{x} \in \mathbb{R}^2 : \left|x_1 - \frac{1}{\lambda}x_2\right| < \frac{M}{\lambda^2}\right\}.$$

Remark 2.3. *For systems having real distinct eigenvalues, we can also derive equivalent bounds on the region of attraction starting from a diagonal state matrix setting. In particular it is very easy to obtain invariant strips for the equivalent system*

$$\dot{\mathbf{z}} = diag(\lambda_1, \lambda_2)\mathbf{z} + \mathbf{H}^{-1}\mathbf{B}u,$$

where

$$\mathbf{H} = \begin{pmatrix} 1 & 1 \\ \lambda_1 & \lambda_2 \end{pmatrix} \quad and\ \mathbf{z} = \mathbf{H}^{-1}\mathbf{x}.$$

Remark 2.4. *The asymptotes of the curves Γ_M and Γ_{-M} are the linear spaces satisfying the equations*

$$x_1 = \frac{1}{\min(\lambda_1, \lambda_2)} x_2 \pm \frac{M}{a_1}.$$

2.2.2 Complex Eigenvalues

This section is devoted to the study of the saturated maximal region of attraction for a system having complex eigenvalues with strictly positive real part. Even if in this case no invariant parallelogram exists, the boundary of the null controllable region is still given by reversed-time system trajectories associated to the saturated control.

Denoting by $\lambda_1 = \alpha + \mathrm{j}\omega$ and $\lambda_2 = \alpha - \mathrm{j}\omega$ the conjugate complex eigenvalues with $\alpha > 0$, the matrix **A** can be written in terms of α and ω as follows

$$\mathbf{A} = \begin{pmatrix} 0 & 1 \\ -\alpha^2 - \omega^2 & 2\alpha \end{pmatrix} \tag{2.9}$$

Without loss of generality we can assume $\omega > 0$. We point out that the approach followed in the case of real eigenvalues fails for the complex case since there is no real solution to (2.8).

Theorem 2.3. *Consider the system (2.5) with state matrix given by (2.9) with $\alpha, \omega > 0$ and assume input saturation (1.3). Setting*

$$s_\pm = \mp \frac{M(1 + \exp(-\alpha\pi/\omega))}{(1 - \exp(-\alpha\pi/\omega))(\alpha^2 + \omega^2)},$$

we denote by $\Psi_{\pm M}$ the trajectories of the solutions $\varphi(t, \mathbf{x}_0, \pm M)$ for $t \in (-\pi/\omega, 0]$ and $\mathbf{x}_0 = (s_\pm, 0)$. The following identity holds

$$\partial \mathscr{B}_M = \Psi_M \cup \Psi_{-M}.$$

Proof. For $u \equiv \pm M$ the solution $\varphi(t, \mathbf{x}_0, \pm M)$ has the following structure

$$\begin{aligned} x_1(t) = (&C_1(\mathbf{x}_0, M)\cos\omega t + \\ &C_2(\mathbf{x}_0, M)\sin\omega t)\exp(\alpha t) \mp \frac{M}{a_1}, \end{aligned}$$

$$\begin{aligned} x_2(t) = (&\alpha(C_1(\mathbf{x}_0, M)\cos\omega t + C_2(\mathbf{x}_0, M)\sin\omega t) + \\ &\omega(-C_1(\mathbf{x}_0, M)\sin\omega t + C_2(\mathbf{x}_0, M)\cos\omega t))\exp(\alpha t), \end{aligned}$$

where C_1, C_2 are real constants depending on $\mathbf{x}_0$ and M.

We see that for any choice of $\mathbf{x}_0$ we have

$$\lim_{t \to -\infty} \varphi(t, \mathbf{x}_0, \pm M) = (\mp \frac{M}{a_1}, 0). \tag{2.10}$$

Let us fix $\mathbf{x}_0 = (s,0)$ with $s \in \mathbb{R}$. We obtain

$$C_1(\mathbf{x}_0,M) = s \pm \frac{M}{a_1}, \qquad C_2(\mathbf{x}_0,M) = -\frac{\alpha}{\omega}C_1.$$

We have

$$x_2(t) = \frac{-\alpha^2-\omega^2}{\omega}\left(s \pm \frac{M}{a_1}\right)\exp(\alpha t)\sin\omega t$$

and we see that $x_2(t) = 0$ for $t = k\pi/\omega$, $k \in \mathbb{Z}$. Setting $\bar{t} = -\pi/\omega$, we have

$$x_1(\bar{t}) = -\left(s \pm \frac{M}{a_1}\right)\exp(-\alpha\pi/\omega) \mp \frac{M}{a_1}.$$

Imposing $x_1(\bar{t}) = -\mathbf{x}_0 = (-s,0)$, we get

$$s = \left(s \pm \frac{M}{a_1}\right)\exp(-\alpha\pi/\omega) \pm \frac{M}{a_1},$$

that is

$$s_\pm = \mp\frac{M(1+\exp(-\alpha\pi/\omega))}{(1-\exp(-\alpha\pi/\omega))(\alpha^2+\omega^2)}.$$

The closed curve $\Psi_M \cup \Psi_{-M}$ is the boundary of a regular domain D. We claim that $D = \mathscr{B}_M$. In particular the closed curves $\Psi_\theta \cup \Psi_{-\theta}$ for $\theta \in [0,M]$ cover the whole domain D and constitute a family of invariant sets for the solution associated to the switching control $u(t) = -\theta S(x_2(t))$, where

$$S : \mathrm{C}^0(\mathbb{R}) \to \{f(t) : f(t) \in \{1,-1\}\ \forall t \in \mathbb{R}\},$$

$$S(g(t)) = \lim_{\tau\to t^-} \mathrm{sign}(g(\tau)).$$

We denote by $Q_{2,4}$ the set

$$Q_{2,4} = \left\{\mathbf{x} \in \mathbb{R}^2 : x_1x_2 < 0\right\}.$$

Let $\mathbf{x}_0$ be an arbitrary point in the inner part of D; by construction there exists $\theta_1 \in (0,M)$ such that $\mathbf{x}_0$ lies on the set $\Psi_{\theta_1} \cup \Psi_{-\theta_1}$. Let us fix $\delta > 0$ such that $\theta_1 + \delta < M$ and apply the control feedback $u_{\delta,\theta_1}(t)$ defined by

$$u_{\delta,\theta_1}(t) = -(\theta_1 + \delta\chi_{\{t\in\mathbb{R}:\mathbf{x}(t)\in Q_{2,4}\}})S(x_2(t)).$$

Recall that for a given set $E \in \mathbb{R}$, the characteristic function $\chi_E = \chi_E(t)$ is defined as follows

$$\chi_E = \begin{cases} 1 & \text{if } t \in E, \\ 0 & \text{if } t \notin E. \end{cases}$$

We denote by t_1 the quantity

$$t_1 = \min\left\{t \in (0,\infty) : \varphi(t,\mathbf{x}_0,u_{\delta,\theta_1}) \in \{x_2 = 0\}\right\};$$

by construction the point $\mathbf{x}_{t_1} = (x_1(t_1),0)$ lies on the closed curve $\Psi_{\theta_2} \cup \Psi_{-\theta_2}$ with $\theta_2 < \theta_1$. Define $t_2 = t_1 + \min\left\{t \in (0,\infty) : \varphi(t,\mathbf{x}_{t_1},u_{\delta_2,\theta_2}) \in \{x_2 = 0\}\right\}$ with $\delta_2 = \delta/2$. In this way we can construct a sequence of parameters $\{\theta_n\} \subset [0,M]$ and a sequence of instants $\{t_n\} \subset (0,\infty)$, $t_n \nearrow +\infty$, such that, if we define

$$u_\delta(t) = \sum_{n=1}^{\infty} u_{\delta_n,\theta_n}(t)\chi_{[t_{n-1},t_n)}, \quad t_0 = 0,\ \delta_n = \delta/n,$$

we have

$$\lim_{t\to\infty} \|\varphi(t,\mathbf{x}_0,u_\delta)\| = 0.$$

We have shown that $D \subseteq \mathscr{B}_M$.

Let us consider now $\mathbf{x}_0 \notin \overline{D}$. Without loss of generality we can assume $\mathbf{x}_0 = (x_1^0,x_2^0)$ with $x_2^0 \leq 0$. We proceed by contradiction supposing the existence of $u(t) \in \mathscr{U}_M$ and $T > 0$ such that $\varphi(T,\mathbf{x}_0,u) \in \Psi_M \cup \Psi_{-M}$. If $\varphi(T,\mathbf{x}_0,u) \in \Psi_M$, since $\dot\varphi(0,\mathbf{x}_0,u) = (x_2^0, a_1x_1^0 + a_2x_2^0 + u(0))$, $\dot\varphi(0,\mathbf{x}_0,M) = (x_2^0, a_1x_1^0 + a_2x_2^0 + M)$, $u(0) \leq M$ and property (2.10) holds, there exists $T_0 \leq T$ such that $\varphi(T_0,\mathbf{x}_0,M) \in \Psi_M$ and this contradicts the invariance of Ψ_M. If $\varphi(T,\mathbf{x}_0,u) \in \Psi_{-M}$, there exists $T_1 < T$ such that $\varphi(T_1,\mathbf{x}_0,u) = (x_1(T_1),x_2(T_1)) = \mathbf{x}_{T_1}$ with $x_2(T_1) \geq 0$ and $\mathbf{x}_{T_1} \notin \Psi_{-M}$. We have $\varphi(T - T_1,\mathbf{x}_{T_1},u) \in \Psi_{-M}$ and $\varphi(T_2,\mathbf{x}_{T_1},-M) \in \Psi_{-M}$ for some $T_2 \leq T - T_1$, that is impossible because of the invariance of Ψ_{-M}. We can conclude that $\mathscr{B}_M \subseteq D$. The claim is proved. □

2.2.3 *Some Remarks*

Remark 2.5. *The sets $\mathscr{B}_{M_1}$ and $\mathscr{B}_{M_2}$ for $M_1 \neq M_2$ are time-uniformly homothetic: there exists a continuous map $\Theta : \mathscr{R}^2 \to \mathscr{R}^2$ such that*

1. *$\Theta(x)$ is an homothety with*

$$\Theta(\mathscr{B}_{M_1}) = \mathscr{B}_{M_2} = \frac{M_2}{M_1}\mathscr{B}_{M_1}.$$

2. *If the solution $\phi(t_0,\mathbf{x}_0,\pm M_1) \in \mathscr{B}_{M_1}$, for some $t_0 \in \mathbb{R}$ and $\mathbf{x}_0 \in \mathscr{B}_{M_1}$, then we have*

$$\Theta(\phi(t_0,\mathbf{x}_0,\pm M_1)) = \phi(t_0, M_2\mathbf{x}_0/M_1, \pm M_2).$$

Remark 2.6. *Consider the system (2.5) and assume that the state matrix* **A** *has real eigenvalues $\lambda_1 = 0$, $\lambda_2 > 0$. For $p,q \in \mathbb{R}$, we define the functions*

$$\psi_{\pm M}(t) = px_1(t) + x_2(t) + q.$$

Following the steps of Section 2.2.1, if we take $p = 0$ and $q = \pm M/\lambda_2$, we obtain

$$\psi_{\pm M}(t) > 0 \Rightarrow \dot{\psi}_{\pm M}(t) > 0$$

$$\psi_{\pm M}(t) < 0 \Rightarrow \dot{\psi}_{\pm M}(t) < 0.$$

We can conclude that the saturated maximal region of attraction for such systems is

$$\mathscr{B}_M = \left\{ \mathbf{x} \in \mathbb{R}^2 : |x_2| < \frac{M}{\lambda_2} \right\}.$$

Remark 2.7. *Consider the system (2.5) and assume the state matrix* **A** *has real positive eigenvalues $\lambda_1, \lambda_2 > 0$. Let us fix the saturation level $M > 0$ and the determinant of the state matrix $-a_1 > 0$; if λ_1 (or λ_2) increases then the saturated maximal region of attraction contracts along the x_2-axis. In particular the widest region is obtained whenever $\lambda_1 = \lambda_2 = \sqrt{-a_1}$.*

Remark 2.8. *Consider the system (2.5) having complex conjugate eigenvalues $\alpha \pm j\omega$. Let us denote with $\mathscr{B}_M^{\alpha,\omega}$ the corresponding saturated maximal region of attraction. If we take $\omega_1, \omega_2 \in \mathbb{R}$ with $\omega_1 > \omega_2 > 0$ then we have*

$$\mathscr{B}_M^{\alpha,\omega_1} \subset \mathscr{B}_M^{\alpha,\omega_2}.$$

In particular the following identity holds

$$\bigcup_{\omega \in (0,\infty)} \mathscr{B}_M^{\alpha,\omega} = \mathscr{B}_M^{\alpha},$$

where $\mathscr{B}_M^{\alpha}$ is the saturated maximal region of attraction of the system (2.5) with real eigenvalues $\lambda_1 = \lambda_2 = \alpha > 0$.

2.3 The SISO n-Dimensional Case

Let us consider a general controllable single-input linear system described by

$$\begin{cases} \dot{\mathbf{x}} = \mathbf{A}\mathbf{x} + \mathbf{B}u \\ \mathbf{x}(0) = \mathbf{x}_0 \end{cases}$$

with $\mathbf{x} \in \mathbb{R}^n$. In view of the controllability assumption, without loss of generality, let us suppose that the pair $(\mathbf{A}, \mathbf{B})$ is given in the Brunovsky canonical form. Denote by $a_1, ..., a_n$ the characteristic coefficients of the state matrix **A**.

The description of the null controllable region has been obtained following several steps; the main ones are described below.

- Invariant strips for the saturated system are found with an algebraic method; the intersection of such strips gives an invariant closed hyperparallelogram, leading to an outer estimate for the null controllable region.
- Any reachable state at time $T > 0$ for an admissible control u is proven to be arbitrarily close to a reachable state at time $T > 0$ for a control input switching a finite number of times between maximum and minimum admissible values.
- A pair of symmetric limit cycles $\Gamma_{\pm M}$ having extrema in two vertices of the invariant hyperparallelogram is considered. The affine hull of such curves is shown to be the whole state space, i.e. there is no affine subspace containing entirely $\Gamma_{\pm M}$.
- The convex hull of $\Gamma_M \cup \Gamma_{-M}$ is a closed set containing the origin as interior point; this set can be enlarged taking for any boundary point the curve corresponding to an extremal trajectory for the reversed-time system. The null controllable region is obtained with an iterative procedure combining convexification and reversed-time evolution.

Let us denote by $\lambda_1,...,\lambda_n \in \mathbb{R}$ the eigenvalues (not necessarily distinct) of the matrix $\mathbf{A}$. Suppose the system is antistable, that is $\lambda_i > 0$ for any $i \in \{1,..,n\}$. Consider the family of surfaces

$$\psi_{\pm M}(t) = x_1(t) + \sum_{i=1}^{n-1} m_i x_{i+1} + q_{\pm M} = 0, \quad m_i, q_{\pm M} \in \mathbb{R}$$

and look for all the possible choices of coefficients m_i in order to have

$$\psi_{\pm M}(t) = 0 \Rightarrow \dot{\psi}_{\pm M}(t) = 0 \text{ for } u \equiv \pm M. \tag{2.11}$$

The derivative of $\psi_{\pm M}(t)$ is given by

$$\dot{\psi}_{\pm M}(t) = x_2 + \sum_{i=1}^{n-2} m_i x_{i+2} + m_{n-1}\left(\pm M + \sum_{i=1}^{n} a_i x_i\right).$$

Using the equality $\psi_{\pm M}(t) = 0$ and imposing $\dot{\psi}_{\pm M}(t) = 0$, the following set of algebraic equations is obtained

$$\begin{cases} q_{\pm M} = \pm M/a_1 \\ 1 + m_{n-1}(-m_1 a_1 + a_2) = 0 \\ m_j + m_{n-1}(-a_1 m_{j+1} + a_{j+2}) = 0, \quad j = 1,...,n-2. \end{cases}$$

Thanks to a recursion property, the system can be rewritten as

$$\begin{cases} m_{n-2} = a_1 m_{n-1}^2 - a_n m_{n-1}, \\ m_{n-3} = a_1^2 m_{n-1}^3 - a_1 a_n m_{n-1}^2 - a_{n-1} m_{n-1}, \\ \vdots \\ m_2 = a_1^{n-3} m_{n-1}^{n-2} - \sum_{j=0}^{n-4} a_1^{n-4-j} a_{n-j} m_{n-1}^{n-3-j}, \\ m_1 = (1 + m_{n-1} a_2)/m_{n-1} a_1, \end{cases}$$

where m_{n-1} satisfies the equation $p_A^*(m_{n-1}) = 0$, with

$$p_A^*(\lambda) = -a_1^{n-1}\lambda^n + 1 + \sum_{j=0}^{n-2} a_1^j a_{j+2}\lambda^{j+1} = 0. \tag{2.12}$$

Proposition 2.2. *Let $\overline{\lambda}$ be a solution of the characteristic equation $p_A(\lambda) = \lambda^n - \sum_{j=1}^n a_j\lambda^{j-1} = 0$. Then $\overline{\lambda}/a_1$ satisfies*

$$p_A^*(\overline{\lambda}/a_1) = 0.$$

Proof. The statement follows observing that

$$\begin{aligned} p_A^*(\overline{\lambda}/a_1) &= -\frac{\overline{\lambda}^n}{a_1} + 1 + \sum_{j=0}^{n-2} a_1^j a_{j+2}\frac{\overline{\lambda}^{j+1}}{a_1^{j+1}} = \\ &= -\frac{1}{a_1}\left(\overline{\lambda}^n - \sum_{j=1}^{n} a_j\overline{\lambda}^{j-1}\right) = 0. \end{aligned}$$ □

As a consequence, the solutions of (2.12) are given by

$$\overline{\lambda}_j = \frac{\lambda_j}{\Pi_{i=1}^n \lambda_i}, \quad j = 1, \ldots, n.$$

The above solutions lead to a set of n pairs of parallel hyperplanes; taking all the possible intersections, a closed hyperparallelogram containing the origin as interior point is found. Denote this set as F_M. Thanks to the invariance property (2.11), if a trajectory starts from a point outside the set F_M, there is no admissible control that can drive it inside, this meaning that

$$\mathscr{B}_M \subseteq F_M.$$

The set ∂F_M intersects the x_1-axis in $\mathbf{x}_0^{\pm} = (\pm M/a_1, 0, \ldots, 0)$. It is immediate to verify that

$$\lim_{t\to-\infty} \phi(t, \mathbf{x}_0^{\pm}, \pm M) = \mathbf{x}_0^{\mp}$$

The following technical lemma is needed.

Lemma 2.1. *Fix $\mathbf{x}_0 \in \mathbb{R}^n$, $u \in \mathscr{U}_M$ and $T > 0$; for any $\varepsilon > 0$ there exists a control function $v_\varepsilon \in \mathscr{U}_M^*$ such that*

1. $\|\phi(T, \mathbf{x}_0, u) - \phi(T, \mathbf{x}_0, v_\varepsilon)\| \leq \varepsilon$;
2. *v_ε has a finite number of switches in $[0, T]$.*

Proof. The statement of the lemma is equivalent to the existence of a function $v_\varepsilon \in \mathscr{U}_M^*$, with a finite number of switches in $[0, T]$, such that

$$J = \left\| \int_0^T e^{\mathbf{A}(T-t)}\mathbf{B}u(t)dt - \int_0^T e^{\mathbf{A}(T-t)}\mathbf{B}v_\varepsilon(t)dt \right\| \leq \varepsilon. \tag{2.13}$$

Consider a general partition $\Pi = \{0 = t_1 < t_2 < ... < t_r = T\}$ of the interval $(0,T)$ and a piecewise constant (PWC) matrix function $\Theta(t)$ associated to it; in particular

$$\Theta(t) = \sum_{i=1}^{r-1} \mathbf{H}_i \chi_{(t_i,t_{i+1})}(t),$$

where $\mathbf{H}_i$ are constant $n \times n$ matrices. The integral in (2.13) can be estimated as

$$\begin{aligned} J \leq & \left\| \int_0^T \left(e^{\mathbf{A}(T-t)} - \Theta(t) \right) \mathbf{B}(u(t) - v_\varepsilon(t)) dt \right\| \\ & + \left\| \int_0^T \Theta(t) \mathbf{B}(u(t) - v_\varepsilon(t)) dt \right\| = J_1 + J_2. \end{aligned}$$

The two terms will be treated separately. The first one verifies

$$\begin{aligned} J_1 & \leq \int_0^T \left\| \left(e^{\mathbf{A}(T-t)} - \Theta(t) \right) \mathbf{B}(u(t) - v_\varepsilon(t)) \right\| dt \\ & \leq 2M \|\mathbf{B}\| \int_0^T \left\| \left(e^{\mathbf{A}(T-t)} - \Theta(t) \right) \right\| dt \end{aligned}$$

for any $v_\varepsilon \in \mathscr{U}_M$. Since $e^{\mathbf{A}(T-t)}$ is continuous, a partition Π_ε with corresponding PWC matrix function Θ_ε can be chosen such that

$$\int_0^T \left\| \left(e^{\mathbf{A}(T-t)} - \Theta_\varepsilon(t) \right) \right\| dt \leq \frac{\varepsilon}{2M\|\mathbf{B}\|}.$$

Let us consider the second term

$$J_2 = \left\| \sum_{i=1}^{r-1} \int_{t_i}^{t_{i+1}} \mathbf{H}_i^\varepsilon \mathbf{B}(u(t) - v_\varepsilon(t)) dt \right\| \leq \sum_{i=1}^{r-1} \left\| \int_{t_i}^{t_{i+1}} \mathbf{H}_i^\varepsilon \mathbf{B}(u(t) - v_\varepsilon(t)) dt \right\|.$$

Since $\mathbf{H}_i^\varepsilon \mathbf{B}$ for any fixed i is a constant vector, the function $v_\varepsilon \in \mathscr{U}_M^*$ can be defined on any subinterval (t_i, t_{i+1}) in order to have $\int_{t_i}^{t_{i+1}} \mathbf{H}_i^\varepsilon \mathbf{B} u(t) dt = \int_{t_i}^{t_{i+1}} \mathbf{H}_i^\varepsilon \mathbf{B} v_\varepsilon(t) dt$. It follows that

$$J_2 = 0.$$

Note that the function v_ε can be designed having at most one switch in any subinterval (t_i, t_{i+1}); this implies that the total number of switches in the whole interval $[0,T]$ is less or equal than $2r < \infty$. □

Proposition 2.3. *Let* $\mathbf{x}_0^\pm = (\pm M/a_1, 0, 0, ..., 0) \in \mathbb{R}^n$ *and consider the curve* $\gamma(t) = \phi(t, \mathbf{x}_0^\pm, \pm M)$ *for* $t \in (-\infty, 0]$. *For* $s < n$ *there is no s-dimensional affine subspace containing* γ.

Proof. A necessary and sufficient condition for a curve to belong to an affine j-dimensional subspace is that its j-curvature is identically null. It will be shown that the $(n-1)$-curvature of $\gamma(t)$ in $\mathbf{x}_0^{\pm}$ is different from zero. With a recursive procedure the derivatives of $\gamma(t)$ can be computed. In particular it holds

$$\gamma^{(1)}(t) = \mathbf{A}\gamma(t) \pm \mathbf{B}M$$

and

$$\gamma^{(j)}(t) = \mathbf{A}\gamma^{(j-1)}(t) = \mathbf{A}^{j-1}\gamma^{(1)}(t) \qquad j = 2, ..., n,$$

where $j \in \mathbb{N}$ denotes the derivative order.

Let us denote by $\mathbf{W}(t)$ the square matrix having $\gamma^{(j)}(t)$, $j = 1, ..., n$, as columns and by $\kappa^{(n-1)}(t)$ the $(n-1)$-curvature of $\gamma(t)$. The condition $\kappa^{(n-1)}(t) \neq 0$ is equivalent to $\det \mathbf{W}(t) \neq 0$. Since in particular

$$\det \mathbf{W}(0) = \pm 2M \neq 0,$$

it has been proved that $\gamma(t)$ is not contained in any $(n-1)$-dimensional affine subspace. □

Let us introduce the set-valued function $\Phi_{\pm M}$ defined on $\mathbb{R}^n$ as follows:

$$\Phi_{\pm M}(\mathbf{x}) = \{\mathbf{y} \in \mathbb{R}^n : \exists t \in (-\infty, 0] : \mathbf{y} = \phi(t, \mathbf{x}, \pm M)\}.$$

In particular the identity $\gamma(t) = \Phi_{\pm M}(\mathbf{x}_0^{\mp})$ holds, where $\mathbf{x}_0^{\pm} = (\pm M/a_1, 0, ..., 0)$.

Corollary 2.1. *Given the set $\Gamma = \Phi_{-M}(\mathbf{x}_0^{-}) \cup \Phi_{M}(\mathbf{x}_0^{+})$ together with its convex hull $S_0 = \mathrm{Co}(\Gamma)$, there exists $\delta > 0$ with $B_\delta(0) \subset S_0$; in particular $|S_0| > 0$.*

Proof. The interior part of $\mathrm{Co}(\Gamma)$ is nontrivial since the affine hull of Γ (i.e. the smallest affine space containing the set Γ) is the whole space $\mathbb{R}^n$ by Proposition 2.3; this proves that $|S_0| > 0$. Let $\bar{\mathbf{x}}$ be an interior point of $\mathrm{Co}(\Gamma)$; by symmetry it follows that $-\bar{\mathbf{x}}$ is still an interior point as well as any point $\bar{\mathbf{x}}_\theta$ obtained as convex combination (see [33] page 45)

$$\bar{\mathbf{x}}_\theta = \theta\bar{\mathbf{x}} - (1-\theta)\bar{\mathbf{x}}, \;\; \theta \in [0,1]. \tag{2.14}$$

The result follows observing that the origin verifies (2.14) with $\theta = \dfrac{1}{2}$. □

The main result of the paper is stated in the next theorem.

Theorem 2.4. *Define a monotone recursive sequence of sets $\{S_j\}_{j \in \mathbb{N}}$*

$$S_{j+1} = \mathrm{Co}(\Phi_{-M}(\partial S_j) \cup \Phi_M(\partial S_j)), \tag{2.15}$$

where S_0 is given by $S_0 = \mathrm{Co}(\Gamma)$ and Γ is defined in Corollary 2.1. The null controllable region of the system is given by

$$\mathscr{B}_M = \lim_{j \to \infty} S_j = \bigcup_{j \in \mathbb{N}} S_j =: S_\infty$$

Proof. First prove that $\mathscr{B}_M \subseteq S_\infty$. Let $\mathbf{x}_0$ be an arbitrary point contained in $\mathscr{B}_M$ and suppose that $\mathbf{x}_0 \notin S_0$; by Corollary 3.6, there exists $\delta > 0$ such that $B_\delta(0) \subset S_0 \subset S_\infty$. Take $0 < \delta_1 < \delta$. Since $\mathbf{x}_0 \in \mathscr{B}_M$, there exist $u \in \mathscr{U}_M$ and $\bar{t} > 0$ such that $\|\phi(t,\mathbf{x}_0,u)\| < \delta_1$ for any $t > \bar{t}$. Thanks to Lemma 2.1 with $\varepsilon = \delta - \delta_1 > 0$, the existence of $v \in \mathscr{U}_M^*$ such that $\|\phi(\bar{t},\mathbf{x}_0,v)\| < \delta$ can be inferred. Moreover, since the trajectory is continuous, it must intersect the boundary ∂S_0; in particular $\phi(\bar{t}_1,\mathbf{x}_0,v) = \tilde{\mathbf{x}} \in \partial S_0$ for some $0 < \bar{t}_1 < \bar{t}$. By construction (see Lemma 2.1) the control $v \in \mathscr{U}_M^*$ has a finite number of switches, say $k \in \mathbb{N}$, as t varies in the bounded interval $[0,\bar{t}_1]$. Consider the reversed-time evolution

$$\phi(t,\tilde{\mathbf{x}},v(t+\bar{t}_1)) \quad \text{for } t \in [-\bar{t}_1,0];$$

it verifies $\phi(-\bar{t}_1,\tilde{\mathbf{x}},v(0)) = \mathbf{x}_0$. Recalling the formula (2.15) for the sets S_j, it follows immediately that $\mathbf{x}_0 \in S_k$, this proving that $\mathbf{x}_0 \in S_\infty$.

Let us show that the converse inclusion $S_\infty \subseteq \mathscr{B}_M$ holds too. Saying that $\mathbf{x}_0 \in S_\infty$ means there exists k such that $\mathbf{x}_0 \in S_k$. Moreover, by definition, there exists a control function in $\mathscr{U}_M^*$ having at most k switches such that the solution driven by it enters S_0 in finite time. It remains only to prove that $S_0 \subseteq \mathscr{B}_M$. Let $\mathbf{x}_0 \in S_0$. Since the solutions are continuously dependant on the system parameters, the interior part of S_0 is completely covered by a family of homothetic closed surfaces Z_{M^*} which are generated starting from the points $\mathbf{x}^{\pm}_{0,M^*} = (\pm M^*/a_1,0,...,0)$ and switching the control between the values $\pm M^*$ as M^* varies in $[0,M]$. Now, if $\mathbf{x}_0^* \in Z_{M^*}$, by the Carathéodory theorem (see [33], page 155), there exist $\theta_j > 0$, $1 < j < n+1$, with $\sum_{j=1}^{n+1} \theta_j = 1$ and $t_j \in (-\infty,0]$ such that

$$\mathbf{x}_0^* = \sum_{j=1}^{j^*} \theta_j \phi(t_j,\mathbf{x}^{+}_{0,M^*},M^*) + \sum_{j=j^*+1}^{n+1} \theta_j \phi(t_j,\mathbf{x}^{-}_{0,M^*},-M^*), \tag{2.16}$$

with $j^* \in [0,n+1]$. Note that the elements of S_∞ obtained as $\lim\limits_{t\to-\infty} \phi(t,\mathbf{x},\pm M)$, for some $\mathbf{x}$, belong to ∂S_∞. It follows from (2.16) that

$$\mathbf{x}^*(t) = \sum_{j=1}^{j^*} \theta_j \phi(t+t_j,\mathbf{x}^{+}_{0,M^*},M^*) + \sum_{j=j^*+1}^{n+1} \theta_j \phi(t+t_j,\mathbf{x}^{-}_{0,M^*},-M^*)$$

satisfies the system equations for initial datum $\mathbf{x}_0^*$ and control input $u^*(t) = M^*$ $(\sum_{j=1}^{j^*} \theta_j - \sum_{j=j^*+1}^{n+1} \theta_j)$. This means that any point in Z_{M^*} lies on a trajectory associated to a suitable admissible control input. Now, if $\mathbf{x}_0 \in S_0$, there exist $M_1^* < M$ such that $\mathbf{x}_0$ lies on the invariant surface $Z_{M_1^*}$ under a suitable control input u^* with $|u^*| \leq M_1^* < M$. In order to stabilize the system, it suffices to make the trajectory jump to a lower level $Z_{M_2^*}$ with $M_2^* < M_1^*$ increasing the control norm. This concludes the proof. □

2.3.1 Remarks and Discussion

It has been clear since earlier works (see [31] for example) that, in order to obtain a good description of null controllable region (maximal recoverable region), it is necessary to deal with time-reversed systems (maximal reachable region). Since the problem has a high computational complexity, the development of different techniques may be useful and interesting. The authors of [24], fixing the saturation level to $M = 1$, describe the boundary of the null controllable region as the boundary of the reachable region for the reversed-time system

$$\partial\mathscr{R} = \left\{ z = \int_{-\infty}^{0} \mathrm{e}^{\mathbf{A}\tau}\mathbf{B}\mathrm{sign}(\mathbf{c}^T \mathrm{e}^{\mathbf{A}\tau}\mathbf{B})d\tau : \mathbf{c} \in \mathbb{R}^n \setminus \{0\} \right\}. \tag{2.17}$$

For systems having only real eigenvalues, the boundary can be described also in terms of trajectories corresponding to control input having at most $n-1$ switches. The result is based on a technical lemma which states the maximum number of solutions for a class of exponential/polynomial algebraic equations.

In this section a different characterization of the null controllable region is presented for antistable systems having real eigenvalues. First an outer estimate is found considering an invariant hyperparallelogram; the second step is to determine a pair of symmetric time-reversed trajectories $\Gamma_{\pm M}$ lying on the boundary of the null controllable region and having extrema in two opposite vertices of the hyperparallelogram. Finally, a bounded domain is obtained taking the convex hull $\mathrm{Co}(\Gamma_M \cup \Gamma_{-M})$. The conclusion follows with an iterative procedure based on a simple algorithm:

1. enlarge the set considering time-reversed evolution starting from boundary points with control input $u \equiv \pm M$;
2. take the convex hull of the new set.

Regarding numerical implementation, an approximation for the null controllable region can be performed by means of Carathéodory theorem with the desired accuracy. Selecting a discrete set of points on $\Gamma_M \cup \Gamma_{-M}$, say $\{\gamma_j\}_{j=1}^{q}$ $q \in \mathbb{N}$, an approximated domain can be computed taking convex combinations of $n+1$ elements :

$$\mathscr{B}_M^{(1)} = \left\{ \mathbf{x} \in \mathbb{R} : \mathbf{x} = \sum_{s=1}^{n+1} \theta_s \gamma_{j_s} : \sum_{s=1}^{n+1} \theta_s = 1 \right\}.$$

The approximation can be further improved repeating the steps of the algorithm described above. In particular fixing a discrete set of negative time instants $\mathscr{T}_q = \{-\infty < t_1 < ... < t_q \leq 0\}$, an increasing sequence of discrete sets can be derived:

$$\mathscr{B}_M^{(j)} = \left\{ \mathbf{x} \in \mathbb{R} : \mathbf{x} = \sum_{s=1}^{n+1} \theta_s \mathbf{z}_s : \sum_{s=1}^{n+1} \theta_s = 1, \mathbf{z}_s \in \Lambda_{j-1} \right\},$$

where $\Lambda_{j-1} = \left\{ \mathbf{z} \in \mathbb{R}^n : \mathbf{z} = \phi(t, \mathbf{x}, \pm M) : t \in \mathscr{T}_q, \mathbf{x} \in \mathscr{B}_M^{(j-1)} \right\}$.

The accuracy of the method is determined by several parameters: the number $q \in \mathbb{N}$, the set $\mathscr{T}_q$, the coefficients θ_s for the convex combinations.

2.4 The MIMO Case

This section is devoted to the presentation of null controllable regions for multi input continuous-time linear systems. The multi input null controllable region can be derived as combination of lower-order single input null controllable regions as well as by a direct computation. Both methods will be presented for sake of completeness.

Let us consider the following multi input linear system

$$\begin{cases} \dot{\mathbf{x}} = \mathbf{A}\mathbf{x} + \mathbf{B}\mathbf{u} \\ \mathbf{x}(0) = \mathbf{x}_0 \end{cases} \tag{2.18}$$

with $m < n$, $\mathbf{A} \in \mathbb{R}^{n \times n}$, $\mathbf{B} \in \mathbb{R}^{n \times m}$, $\mathbf{x} \in \mathbb{R}^n$ and $\mathbf{u} \in \mathbb{R}^m$; the plant is assumed to be controllable, i.e.

$$\text{rank}[\mathbf{B}\ \mathbf{A}\mathbf{B}\ \mathbf{A}^2\mathbf{B}\ \cdots\ \mathbf{A}^{n-1}\mathbf{B}] = n \tag{2.19}$$

and antistable, that is $\Re(\lambda_i) > 0$ for any eigenvalue λ_i, $i = 1, ..., n$, of the matrix $\mathbf{A}$. We will use the following notation:

$$\mathbf{B} = [\mathbf{B}_1\ \mathbf{B}_2\ \cdots\ \mathbf{B}_m], \quad \mathbf{u} = \begin{bmatrix} u_1 \\ u_2 \\ \vdots \\ u_m \end{bmatrix},$$

where $\mathbf{B}_i \in \mathbb{R}^n$ and $u_i \in \mathbb{R}$ for any $i = 1, ..., m$.

2.4.1 Method of Lower-Order Single Input Subsystems

According to (2.18), we can define a family of single input systems Σ_i, $i = 1, .., m$, as follows

$$\Sigma_i = \begin{cases} \dot{\mathbf{x}} = \mathbf{A}\mathbf{x} + \mathbf{B}_i u_i \\ \mathbf{x}(0) = \mathbf{x}_0 \end{cases}$$

Let $n_i \in \mathbb{N}$ be the dimension of the controllability subspace for the system Σ_i,

$$n_i := \text{rank}[\mathbf{B}_i\ \mathbf{A}\mathbf{B}_i\ \mathbf{A}^2\mathbf{B}_i\ \cdots\ \mathbf{A}^{n-1}\mathbf{B}_i] \leq n.$$

The system Σ_i can be transformed into a controllable/uncontrollable subsystems decomposition by a linear coordinates transformation; let $\mathbf{R}_i \in \mathbb{R}^{n\times n}$ the matrix associated to such transformation. We have

$$\mathbf{R}_i^{-1}\mathbf{A}\mathbf{R}_i = \begin{bmatrix} \mathbf{A}_c^{(i)} & \mathbf{0}^{n_i\times(n-n_i)} \\ \mathbf{0}^{(n-n_i)\times n} & \mathbf{A}_{uc}^{(i)} \end{bmatrix}, \quad \mathbf{R}_i^{-1}\mathbf{B}_i = \begin{bmatrix} \mathbf{B}_i^* \\ 0 \\ \vdots \\ 0 \end{bmatrix}$$

with $\mathbf{A}_c^{(i)} \in \mathbb{R}^{n_i\times n_i}$, $\mathbf{A}_{uc}^{(i)} \in \mathbb{R}^{(n-n_i)\times(n-n_i)}$, $\mathbf{B}_i^* \in \mathrm{R}^{n_i}$ and

$$\operatorname{rank}\left[\mathbf{B}_i^* \; \mathbf{A}_c^{(i)}\mathbf{B}_i^* \; \left(\mathbf{A}_c^{(i)}\right)^2 \mathbf{B}_i^* \cdots \left(\mathbf{A}_c^{(i)}\right)^{n_i-1} \mathbf{B}_i^*\right] = n_i.$$

As a consequence, the lower-order system $(\mathbf{A}_c^{(i)}, \mathbf{B}_i^*)$ can be transformed in its controllability canonical form by a linear transformation $\mathbf{Q}_i \in \mathrm{R}^{n_i\times n_i}$,

$$\mathbf{Q}_i^{-1}\mathbf{A}_c^{(i)}\mathbf{Q}_i = \bar{\mathbf{A}}_i := \begin{bmatrix} 0 & 1 & 0 & \cdots & 0 \\ 0 & 0 & 1 & \cdots & 0 \\ \vdots & \vdots & \vdots & \ddots & \vdots \\ 0 & 0 & 0 & \vdots & 1 \\ a_1^{(i)} & a_2^{(i)} & a_3^{(i)} & \cdots & a_{n_i}^{(i)} \end{bmatrix}, \quad \mathbf{Q}_i^{-1}\mathbf{B}_i^* = \bar{\mathbf{B}}_i := \begin{bmatrix} 0 \\ 0 \\ \vdots \\ 0 \\ 1 \end{bmatrix}.$$

Using the results of Section 2.3, we can compute the null controllable region for the system $(\bar{\mathbf{A}}_i, \bar{\mathbf{B}}_i)$; let us denote this region by $\mathscr{B}_M^{(i)}$. The description of $\mathscr{B}_M^{(i)}$ with respect to the original coordinates of the system Σ_i is given by the inverse transformation

$$\mathbf{R}_i\mathbf{H}_i\left(\mathscr{B}_M^{(i)} \times \{\mathbf{0}^{n-n_i}\}\right) := \mathscr{D}_M^{(i)},$$

where the matrix $\mathbf{H}_i \in \mathbb{R}^{n\times n}$ is defined as

$$\mathbf{H}_i := \begin{bmatrix} \mathbf{Q}_i & \mathbf{0}^{n_i\times(n-n_i)} \\ \mathbf{0}^{(n-n_i)\times n} & \mathbf{I}_{(n-n_i)} \end{bmatrix}.$$

By construction, the set $\mathscr{D}_M^{(i)}$ is a convex set contained in a linear subspace $V_i \subset \mathbb{R}^n$ with $\dim(V_i) = n_i$.

Theorem 2.5. *The null controllable region $\mathscr{B}_M$ for the multi input system (2.18) is given by*

$$\mathscr{B}_M = \sum_{i=1}^{m} \mathscr{D}_M^{(i)} = \left\{\mathbf{x} \in \mathbb{R}^n : \mathbf{x} = \mathbf{x}_1 + \mathbf{x}_2 + \cdots + \mathbf{x}_m,\ \mathbf{x}_i \in \mathscr{D}_M^{(i)}\ \forall\, i = 1,...,m\right\}.$$

Proof. The set $\sum_{i=1}^m \mathscr{D}_M^{(i)}$, as it is defined as a finite sum of convex sets, is still convex (see for instance [33]). Let $\mathbf{x} \in \sum_{i=1}^m \mathscr{D}_M^{(i)}$, with $\mathbf{x} = \sum_{i=1}^m \mathbf{x}_i$. Since $\mathbf{x}_i \in \mathscr{D}_M^{(i)}$, by construction, there exists $u_i \in \mathscr{U}_M$ such that, defining $\phi_i(t, \mathbf{x}_0, u)$ the dynamics of the system Σ_i driven by the control u, we have

$$\lim_{t \to \infty} ||\phi_i(t, \mathbf{x}_i, u_i)|| = 0.$$

Now, since

$$\phi(t, \mathbf{x}, \mathbf{u}) = \sum_{i=1}^m \phi_i(t, \mathbf{x}_i, u_i),$$

we can deduce the existence of $\mathbf{u} \in \mathscr{U}_\mathbf{M}$ such that

$$\lim_{t \to \infty} ||\phi_i(t, \mathbf{x}, \mathbf{u})|| = 0;$$

in other words we have $\sum_{i=1}^m \mathscr{D}_M^{(i)} \subseteq \mathscr{B}_M$. Let us show that the converse inclusion $\mathscr{B}_M \subseteq \sum_{i=1}^m \mathscr{D}_M^{(i)}$ holds too. Suppose that $\mathbf{x} \in \mathscr{B}_M$; by definition, there exists $\mathbf{u} = (u_1, ..., u_m)$ with $|u_i| \leq M$ such that

$$\lim_{t \to \infty} ||\phi_i(t, \mathbf{x}, \mathbf{u})|| = 0. \tag{2.20}$$

Using the explicit formula for the solution we have

$$\phi_i(t, \mathbf{x}, \mathbf{u}) = \mathrm{e}^{\mathbf{A}t}\mathbf{x} + \int_0^t \mathrm{e}^{\mathbf{A}(t-s)} \sum_{i=1}^m \mathbf{B}_i u_i(s) ds.$$

From (2.20) we obtain

$$\lim_{t \to \infty} \mathrm{e}^{\mathbf{A}t} \left(\mathbf{x} + \mathrm{e}^{\ \mathbf{A}t} \int_0^t \mathrm{e}^{\mathbf{A}(t-s)} \sum_{i=1}^m \mathbf{B}_i u_i(s) ds \right) = 0$$

and necessarily one has

$$\lim_{t \to \infty} \mathrm{e}^{-\mathbf{A}t} \int_0^t \mathrm{e}^{\mathbf{A}(t-s)} \sum_{i=1}^m \mathbf{B}_i u_i(s) ds = \lim_{t \to \infty} \int_0^t \mathrm{e}^{-\mathbf{A}s} \sum_{i=1}^m \mathbf{B}_i u_i(s) ds = -\mathbf{x}.$$

Now, since $|| \sum_{i=1}^m \mathbf{B}_i u_i(s) || \leq M \sum_{i=1}^m ||\mathbf{B}_i|| < \infty$ and $\int_0^\infty ||\mathrm{e}^{-\mathbf{A}s} ds|| < \infty$, the following limits exist $\forall i = 1, ..., m$ and they are finite:

$$\lim_{t \to \infty} \int_0^t \mathrm{e}^{-\mathbf{A}s} \mathbf{B}_i u_i(s) ds := -\mathbf{x}_i.$$

Moreover, to ensure the asymptotic stability of the system, the input control $\mathbf{u} = (u_1, u_2, ..., u_m)$ has to satisfy

$$\lim_{t \to \infty} |u_1(s)| = \lim_{t \to \infty} |u_2(s)| = \cdots = \lim_{t \to \infty} |u_m(s)| = 0$$

and, as a consequence, the following property holds

$$\lim_{t\to\infty} e^{\mathbf{A}t}\left(\mathbf{x}_i + \int_0^t e^{-\mathbf{A}s}\mathbf{B}_i u_i(s)ds\right) = 0 \; \forall i = 1,...,m.$$

In conclusion, we have $\mathbf{x}_i \in \mathscr{D}_M^{(i)}$ and $\mathbf{x} = \sum_{i=1}^m \mathbf{x}_i$, that is $\mathscr{B}_M \subseteq \sum_{i=1}^m \mathscr{D}_M^{(i)}$. The proof is completed. □

2.4.2 Direct Method

The direct method for the description of the multi input null controllable region follows the same steps presented in the single input case. In particular, the null controllable region can be computed with an iterative procedure combining reversed-time evolution with extremal controls and set convexification.

The following result, which constitutes the multi input version of Lemma 2.1, can be proved.

Lemma 2.2. *Fix* $\mathbf{x}_0$ *and* $\mathbf{u} = (u_1,...,u_m) \in \mathscr{U}_\mathbf{M}$*; then for every* $T > 0$ *and for any* $\varepsilon > 0$*, there exists* $\mathbf{v}^{(\varepsilon)} \in \mathscr{U}_\mathbf{M}^*$ *such that*

$$||\phi(T,\mathbf{x}_0,\mathbf{u}) - \phi(T,\mathbf{x}_0,\mathbf{v}^{(\varepsilon)})|| \le \varepsilon.$$

Moreover each component of $\mathbf{v}^{(\varepsilon)}$ *has a finite number of switches in the interval* $[0,T]$.

Proof. Using the explicit formula for the solution, the statement of the lemma is equivalent to the existence of a bang-bang control $\mathbf{v}^{(\varepsilon)} \in \mathscr{U}_\mathbf{M}^*$ with a finite number of switchings such that

$$J^\varepsilon := \left\| \int_0^T e^{A(T-s)}\mathbf{B}(\mathbf{u}(s) - \mathbf{v}^{(\varepsilon)}(s))ds \right\| \le \varepsilon.$$

Consider now a general partition $\Pi_r = \{0 = t_1 < t_2 < ... < t_r = T\}$ of the interval $(0,T)$ and define a piecewise constant matrix (PWC) function $\Theta(t)$ associated to it:

$$\Theta(t) = \sum_{k=1}^{r-1} \mathbf{H}_k \chi_{(t_k,t_{k+1})}(t), \tag{2.21}$$

where $\mathbf{H}_i$ are constant $n \times m$ matrices to be fixed and $\chi_{(t_k,t_{k+1})}(t)$ are standard characteristic functions. Setting

$$\mathscr{H}(t) := e^{\mathbf{A}(t-s)}\mathbf{B} - \Theta(t),$$

the integral J^{ε} can be estimates as

$$J^{\varepsilon} \leq \left\| \int_0^T \mathscr{H}(s)(\mathbf{u}(s) - \mathbf{v}^{(\varepsilon)}(s))ds \right\| + \left\| \int_0^T \Theta(s)(\mathbf{u}(s) - \mathbf{v}^{(\varepsilon)}(s))ds \right\| =: J_1^{\varepsilon} + J_2^{\varepsilon}.$$

The two terms will be treated separately. The first one verifies

$$J_1^{\varepsilon} \leq \int_0^T \left\| \mathscr{H}(s)(\mathbf{u}(s) - \mathbf{v}^{(\varepsilon)}(s)) \right\| ds \leq 2M \int_0^T \|\mathscr{H}(s)\| \, ds$$

for any $\mathbf{v}^{(\varepsilon)} \in \mathscr{U}_{\mathbf{M}}^*$. Since the matrix $e^{\mathbf{A}(t-s)}\mathbf{B} \in L^1_{loc}(\mathbb{R};\mathbb{R}^{n\times m})$ and the PWC functions defined in (2.21) are dense with respect to L^1-norm (see [34]), suitable matrices $\mathbf{H}_k^{\varepsilon}$ and a partition Π_{r_ε} can be found such that

$$\int_0^T \|\mathscr{H}(s)\| \, ds \leq \frac{\varepsilon}{2M}. \tag{2.22}$$

Let us consider the second term

$$J_2^{\varepsilon} = \left\| \sum_{k=1}^{r_\varepsilon - 1} \int_{t_k}^{t_{k+1}} \mathbf{H}_k^{\varepsilon}(\mathbf{u}(s) - \mathbf{v}^{(\varepsilon)}(s))ds \right\| \leq \sum_{k=1}^{r_\varepsilon - 1} \left\| \int_{t_k}^{t_{k+1}} \mathbf{H}_k^{\varepsilon}(\mathbf{u}(s) - \mathbf{v}^{(\varepsilon)}(s))ds \right\|.$$

Since $\mathbf{H}_k^{\varepsilon}$ for any fixed k is a constant matrix, the control $\mathbf{v}^{\varepsilon}$ can be defined on any subinterval (t_k, t_{k+1}) in order to have

$$\int_{t_k}^{t_{k+1}} \mathbf{u}(s)ds = \int_{t_k}^{t_{k+1}} \mathbf{v}^{(\varepsilon)}(s)ds;$$

this can be done following the simple rule

$$v_i^{(\varepsilon)}(t)|_{(t_k,t_{k+1})} := \begin{cases} M & \text{if} \quad t_k \leq t \leq \tau_k^i \\ -M & \text{if} \quad \tau_k^i < t \leq t_{k+1}, \end{cases}$$

for $i = 1,...,m$ where $\tau_k^i = \dfrac{\int_{t_k}^{t_{k+1}} u_i(s)ds + M(t_{k+1}+t_k)}{2M}$. It follows that, with the above choice for the components of the bang-bang control $v^{(\varepsilon)}$, the integral J_2^ε verifies

$$J_2^\varepsilon \leq \sum_{k=1}^{r_\varepsilon-1} \left\| \mathbf{H}_k^\varepsilon \int_{t_k}^{t_{k+1}} (\mathbf{u}(s) - \mathbf{v}^{(\varepsilon)}(s))ds \right\| = 0.$$

Note that in any subinterval (t_k, t_{k+1}) the functions $v_i^{(\varepsilon)}$ have at most one switch by construction; this implies that the total number of switches in the whole interval $[0,T]$ is less or equal than $2(r_\varepsilon - 1)$. □

Let us denote by $\mathscr{U}_M^{**}$ the set of extremal controls; to this purpose we recall that a control input $\mathbf{u} = (u_1,...,u_m) \in \mathscr{U}_M^*$ is extremal if $u_i \equiv \pm M\ \forall i = 1,...,m$. Using the explicit formula for the solution $\phi(t,\mathbf{x},\mathbf{u})$ one can verify that $\forall \mathbf{u} \in \mathscr{U}_M^{**}$

$$\lim_{t\to-\infty} \phi(t,\mathbf{x},\mathbf{u}^*) = \mathbf{A}^{-1}\mathbf{B}\mathbf{u}^*\ \forall \mathbf{x} \in \mathbb{R}^n.$$

Fixing $u_1^* \equiv M$, there are 2^{m-1} possible combinations for the other components of an extremal control $\mathbf{u} \in \mathscr{U}_M^{**}$; in this way, we can define a set of 2^{m-1} points

$$\mathbf{x}_{M,\mathbf{u}} = \mathbf{A}^{-1}\left(\mathbf{B}_1 + \sum_{i=2}^{m} \pm\mathbf{B}_i\right)M$$

such that

$$\lim_{t\to-\infty} \phi(t, \mp\mathbf{x}_{M,\mathbf{u}}, \pm\mathbf{u}) = \pm\mathbf{x}_{M,\mathbf{u}}\ \forall \mathbf{u} \in \mathscr{U}_{M,+}^{**}$$

where it has been set $\mathscr{U}_{M,+}^{**} := \{\mathbf{u} \in \mathscr{U}_M^{**} : u_1 = M\}$. For $\mathbf{u} \in \mathscr{U}_{M,+}^{**}$ we define the family of curves

$$\gamma_{M,\mathbf{u}}(t) := \phi(t, -\mathbf{x}_{M,\mathbf{u}}, \mathbf{u}),\ t \in (-\infty, 0]$$

and the corresponding supports

$$\Gamma_{M,\mathbf{u}} := \{\mathbf{x} \in \mathbb{R}^n : \mathbf{x} = \phi(t, -\mathbf{x}_{M,\mathbf{u}}, \mathbf{u}),\ t \in (-\infty, 0]\}. \tag{2.23}$$

Proposition 2.1. *There exists at least one extremal control $\bar{\mathbf{u}} \in \mathscr{U}_{M,+}^{**}$ such that, for $s < n$ there is no s-dimensional affine subspace containing $\gamma_{M,\bar{\mathbf{u}}}(t)$.*

Proof. For any fixed $\mathbf{u} \in \mathscr{U}_{M,+}^{**}$ the multi input system $(\mathbf{A},\mathbf{B})$ can be regarded as the single input system $(\mathbf{A},\mathbf{B}_\mathbf{u})$ driven by $u \equiv M$ with

$$\mathbf{B}_\mathbf{u} = \mathbf{B}_1 + \sum_{i=1}^{m} \pm\mathbf{B}_i; \tag{2.24}$$

the signs in the above formula are fixed according to the signs of the components u_i, $i = 2,...m$. Since the plant $(\mathbf{A},\mathbf{B})$ is assumed to be controllable, there exists necessarily $\bar{\mathbf{u}} \in \mathscr{U}^{**}_{M,+}$ such that the single input system $(\mathbf{A},\mathbf{B}_{\bar{\mathbf{u}}})$, with $\mathbf{B}_{\bar{\mathbf{u}}}$ defined by (2.24), is controllable as well. Now the curve $\gamma_{M,\bar{\mathbf{u}}}(t)$ can be regarded as the reversed-time evolution of the single input system $(\mathbf{A},\mathbf{B}_{\bar{\mathbf{u}}})$ driven by the control $u \equiv M$ and the conclusion follows applying the results given in Proposition 2.3 to such curve. □

Introducing the set-valued function $\Phi_{\pm\mathbf{u}}$, $\mathbf{u} \in \mathscr{U}^{**}_{M,+}$, defined on $\mathbb{R}^n$ as follows

$$\Phi_{\mathbf{u}}(\mathbf{x}) = \{\mathbf{y} \in \mathbb{R}^n : \exists t \in (-\infty,0] : \mathbf{y} = \phi(t,\mathbf{x},\pm\mathbf{u})\},$$

one can design a monotone recursive sequence of sets $\{S_j\}_{j\in\mathbb{N}}$

$$S_{j+1} = \mathrm{Co}\left(\bigcup_{\mathbf{u}\mathscr{U}^{**}_{M,+}} (\Phi_{-\mathbf{u}}(\partial S_j) \cup \Phi_{\mathbf{u}}(\partial S_j))\right), \tag{2.25}$$

where S_0 is given by $S_0 = \mathrm{Co}\left(\bigcup_{\mathbf{u}\mathscr{U}^{**}_{M,+}} -\Gamma_{M,\mathbf{u}} \cup \Gamma_{M,\mathbf{u}}\right)$ with $\Gamma_{M,\mathbf{u}}$ given by (2.23).

Theorem 2.6. *The null controllable region for the multi input system (2.18) is*

$$\mathscr{B}_M = \lim_{j\to\infty} S_j = \bigcup_{j\in\mathbb{N}} S_j =: S_\infty$$

Proof. Using Lemma 2.2, the dynamics of the system driven by an admissible control can be approximated with the desired accuracy by the dynamics corresponding to extremal controls; due to this result, the arguments to prove the theorem are analogous to those employed to demonstrate the validity of Theorem 2.4 and for sake of brevity the detailed proof is omitted. □

2.5 Examples

Example 2.1. We consider a planar system having real eigenvalues $\lambda_1 = 1$ and $\lambda_2 = 3$, therefore the state matrix is

$$\mathbf{A} = \begin{pmatrix} 0 & 1 \\ -3 & 4 \end{pmatrix}$$

Fixing the saturation level to $M = 4$, the equations of $\Gamma_M = -\Gamma_{-M}$ are

$$\Gamma_M = \begin{cases} x_1(t) = \dfrac{2}{3}(e^t-1)^2(e^t+2) + \dfrac{2}{3}e^t(e^{2t}-3), \\ x_2(t) = 4e^t(e^{2t}-1), \qquad t \in (-\infty,0] \end{cases}$$

Figure 2.1 shows a plot of the saturated maximal region of attraction.

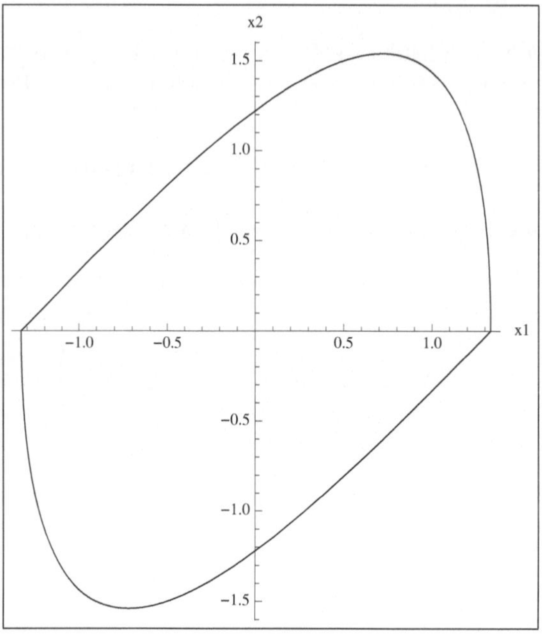

Fig. 2.1. Example of saturated maximal region of attraction with $\lambda_1 = 1,\ \lambda_2 = 3$ and $M = 4$.

Example 2.2. Let us consider a planar systems with complex eigenvalues $\lambda_1, \lambda_2 \in \mathbb{C}$ with $\Re(\lambda_1) = \Re(\lambda_2) = 0.5$ and $\Im(\lambda_1) = -\Im(\lambda_2) = 1.2$. The canonical form of the state matrix is

$$\mathbf{A} = \begin{pmatrix} 0 & 1 \\ -1.69 & 1 \end{pmatrix}.$$

Fixing the saturation level to $M = 2$, the equations of $\Psi_M = -\Psi_{-M}$ are

$$\Psi_M = \begin{cases} x_1(t) = \dfrac{100}{507}\left(6 + \dfrac{e^{\frac{5\pi}{12} + \frac{t}{2}}(-12\cos\frac{6t}{5} + 5\sin\frac{6t}{5})}{e^{\frac{5\pi}{12}} - 1}\right), \\[2ex] x_2(t) = \dfrac{10e^{\frac{5\pi}{12} + \frac{t}{2}}\sin\frac{6t}{5}}{3(e^{\frac{5\pi}{12}} - 1)} \end{cases}$$

The null controllable region for the system is depicted in Fig.2.2.

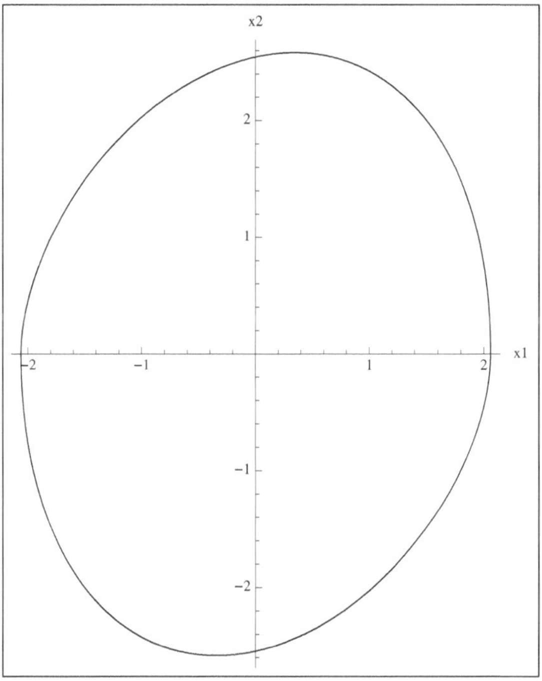

Fig. 2.2. Example of saturated maximal region of attraction with $\Re(\lambda_1) = \Re(\lambda_2) = 1.5$, $\Im(\lambda_1) = -\Im(\lambda_2) = 1.2$ and $M = 2$.

Example 2.3. Let us consider the antistable system $(\mathbf{A}, \mathbf{B})$ with

$$\mathbf{A} = \begin{pmatrix} 0 & 1 & 0 \\ 0 & 0 & 1 \\ 0.1 & -0.8 & 1.7 \end{pmatrix}, \qquad \mathbf{B} = \begin{pmatrix} 0 \\ 0 \\ 1 \end{pmatrix}.$$

The open-loop system eigenvalues are

$$\lambda_1 = 0.2, \quad \lambda_2 = 0.5, \quad \lambda_3 = 1.$$

Setting the saturation level $M = 1$, the starting points for the algorithm are given by

$$\mathbf{x}_0^{\pm} = \pm(10, 0, 0).$$

Pictures of the null controllable region at different iteration steps are given in Fig.2.3. The simulation has been performed prescribing symmetric convex combinations, i.e. $\theta_1 = \theta_2 = \cdots = \theta_{N+1} = 1/(N+1)$ and using the `convhulln` MatLab built-in function.

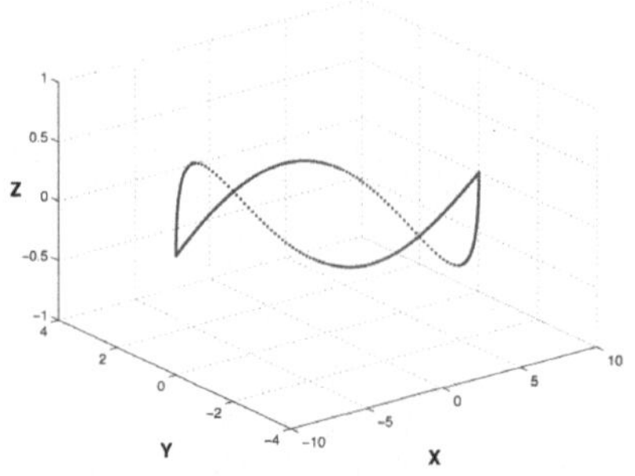

(a) - Curve Γ

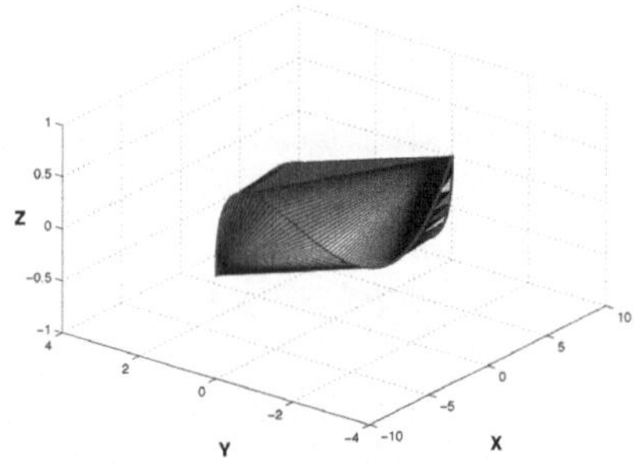

(b) - Initial approximated set S_0

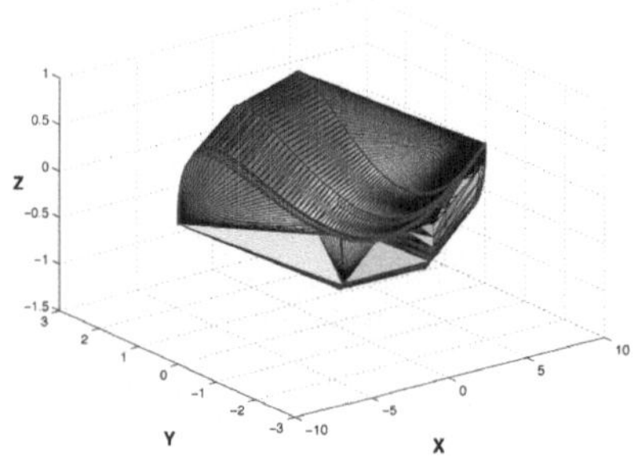

(c) - Approximation after 10 iterations

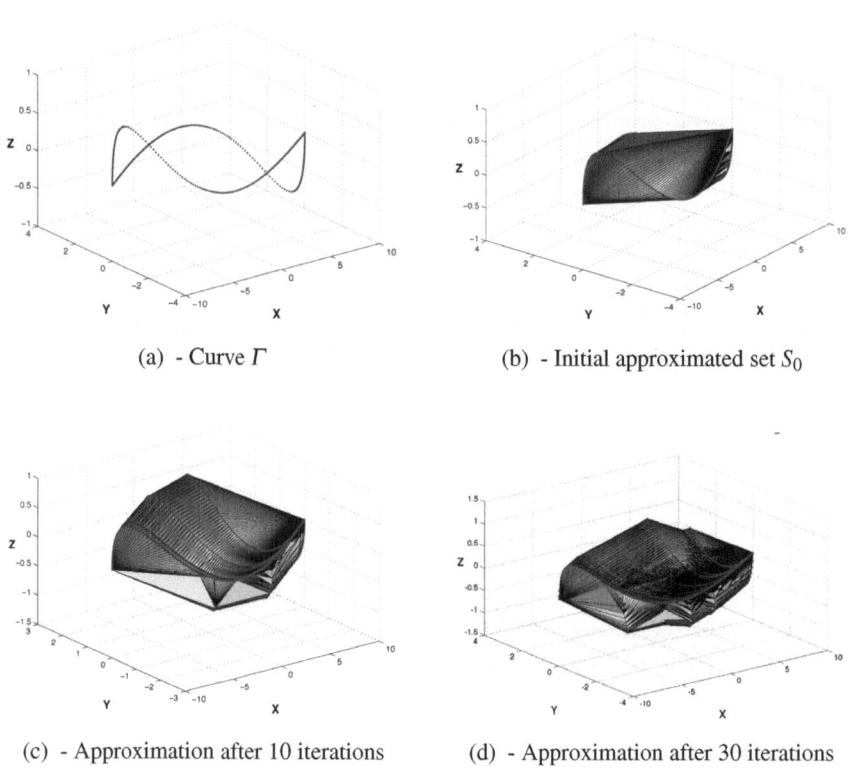

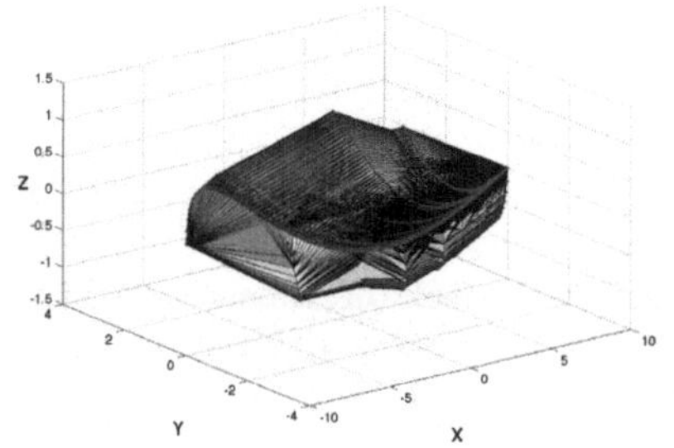

(d) - Approximation after 30 iterations

Fig. 2.3. Approximation of the null controllable region

Chapter 3
Estimation of the Null Controllable Region: Discrete-Time Plants

This chapter is devoted to the study of the null controllable region for a saturating discrete-time plant. In the SISO case, analogously to the continuos-time case, the null controllable region is characterized making resort to a recursive formula defining sets whose union completely describes it, wherever are located the complex eigenvalues of the matrix **A** with respect to the unit circle.

The following notations will be used in the following. Denote by $\ell^\infty(\mathbb{Z})$ the set of bounded bilateral sequences. For any arbitrary control input $u \in \ell^\infty(\mathbb{Z}$ and a fixed initial condition $\mathbf{x}_0$, let us denote by $\phi(k, \mathbf{x}_0, u)$ the solution of the system

$$\begin{cases} \mathbf{x}(k+1) = \mathbf{A}\mathbf{x}(k) + \mathbf{B}u(k) \\ \mathbf{x}(0) = \mathbf{x}_0 \end{cases} \tag{3.1}$$

with $\mathbf{x} = (x_1, ..., x_n) \in \mathbb{R}^n$. In view of the controllability assumption, without loss of generality, the pair $(\mathbf{A}, \mathbf{B})$ is given in canonical form. Given a real positive value $M > 0$, the system is said to be subject to saturation at level M if the control variable $u(k)$ is forced to have norm less than M. By the notation $\mathscr{W}_M$ will indicate the set of admissible control functions:

$$\mathscr{W}_M = \left\{ u \in \ell^\infty(\mathbb{Z}) : \|u\|_\infty = \sup_{k \in \mathbb{Z}} |u(k)| \leq M \right\}.$$

As in the continuous case the null controllable region $\mathscr{B}_M^\sharp$ is defined as the maximal region of attraction for the system under the input constraint, i.e.

$$\mathscr{B}_M^\sharp = \left\{ \mathbf{x} \in \mathbb{R}^n : \exists u \in \mathscr{W}_M : \lim_{k \to +\infty} \phi(k, \mathbf{x}, u) = 0 \right\}.$$

The set $\mathscr{W}_M^* \subset \mathscr{W}_M$ is the set of admissible switching control functions

$$\mathscr{W}_M^* = \{ u \in \mathscr{W}_M : u(k) \in \{M, -M\} \ \forall k \in \mathbb{Z} \}.$$

M.L. Corradini et al.: Control Systems with Saturating Inputs, LNCIS 424, pp. 33–52.
springerlink.com

It is worth noting that, albeit an algorithm describing the null controllable region has been already given in [35], the approach described in the present chapter is different and follows the lines of the procedure used in the continuous case. As in the previous chapter, the SISO planar case is treated first, then the null controllable region for multidimensional systems having positive real eigenvalues with norm larger than 1 is described. Finally the MIMO case is addressed in the last section.

3.1 The SISO Planar Case

Let us consider the following planar single-input discrete-time linear system

$$\begin{cases} \mathbf{x}(k+1) = \mathbf{A}\mathbf{x}(k) + \mathbf{B}u(k) \\ \mathbf{x}(0) = \mathbf{x}_0 \end{cases} \tag{3.2}$$

with $\mathbf{x} = (x_1, x_2) \in \mathbb{R}^2$. In view of the controllability assumption, without loss of generality, the pair $(\mathbf{A}, \mathbf{B})$ is given in canonical form. The system is assumed to be subject to input saturation, i.e.

$$u(k) = \text{sat}_M(v(k)).$$

The elements a_1, a_2 on the last row of the matrix $\mathbf{A}$ are the coefficients of the characteristic polynomial, i.e. $p_{\mathbf{A}}(\lambda) = -\lambda^2 + a_2\lambda + a_1$. The null controllable region $\mathscr{B}_M^\sharp$ in this case is defined as

$$\mathscr{B}_M^\sharp = \left\{ \mathbf{x} \in \mathbb{R}^2 : \exists u \in \mathscr{W}_M : \lim_{k \to +\infty} \phi(k, \mathbf{x}, u) = 0 \right\}$$

Let us define the functions

$$\psi_{\pm M}(t) = x_1(t) + m x_2(t) + q;$$

our aim is to determine constants $m, q \in \mathbb{R}$ such that, when the control variable u is taken identically equal to $\pm M$, we have

$$\psi_{\pm M}(t) = 0 \Rightarrow \psi_{\pm M}(t+1) = 0.$$

We have

$$\begin{aligned} \psi_{\pm M}(t+1) &= x_1(t+1) + m x_2(t+1) + q \\ &= x_2(t) + m(a_1 x_1(t) + a_2 x_2(t) \pm M) + q; \end{aligned} \tag{3.3}$$

imposing $\psi_{\pm M}(t) = 0$ we get $x_1(t) = -m x_2(t) - q$. The equation $\psi_{\pm M}(t+1) = 0$ can be written as

$$\psi_{\pm M}(t+1) = x_2(t) + m(-a_1 m x_2(t) - a_1 q + a_2 x_2(t) \pm M) + q = 0.$$

We obtain

$$\begin{cases} q = \pm \dfrac{mM}{a_1 m - 1} \\ a_1 m^2 - a_2 m - 1 = 0. \end{cases}$$

From the last equation we get

$$m = \frac{a_2 \pm \sqrt{a_2^2 + 4a_1}}{2a_1}; \tag{3.4}$$

the expression $a_2^2 + 4a_1$ is non negative if and only if A has real eigenvalues. In particular we have

$$m = \frac{\lambda_1 + \lambda_2 \pm |\lambda_1 - \lambda_2|}{-2\lambda_1\lambda_2} \in \left\{ -\frac{1}{\lambda_1}, -\frac{1}{\lambda_2} \right\}.$$

We denote by E_M^1, E_M^2 the (unbounded) strips

$$E_M^1 = \left\{ \mathbf{x} \in \mathbb{R}^2 : \left| x_1 - \frac{1}{\lambda_1} x_2 \right| < \frac{M}{|\lambda_1\lambda_2 - \lambda_1|} \right\}$$

$$E_M^2 = \left\{ \mathbf{x} \in \mathbb{R}^2 : \left| x_1 - \frac{1}{\lambda_2} x_2 \right| < \frac{M}{|\lambda_1\lambda_2 - \lambda_2|} \right\}.$$

If it is assumed that $|\lambda_1|, |\lambda_2| > 1$, as a consequence of the previous analysis, the system cannot be driven inside the strips by a control input satisfying the constraint $|u(k)| < M$ if the initial condition is taken outside the set E_M^1 or the set E_M^2. In other words, the states belonging to the complementary sets $\mathbb{R}^2 \setminus E_M^1$ and $\mathbb{R}^2 \setminus E_M^2$ cannot be steered to the origin under any admissible control $-M \le u(k) \le M$, this leading to an outer estimate for the maximal region of attraction, as described in the next statement.

Proposition 3.1. *If the eigenvalues λ_1, λ_2 of the state matrix $\mathbf{A}$ are real numbers placed outside the unit circle, i.e. $|\lambda_1|, |\lambda_2| > 1$, then the following set inclusion holds*

$$\mathscr{B}_M^{\sharp} \subset E_M^1 \cap E_M^2. \tag{3.5}$$

The boundary sets ∂E_M^1 and ∂E_M^2 have intersection along the main bisector $x_1 = x_2$ and such intersection is given by a pair of symmetric points

$$\mathbf{x}_0^{\pm} = \pm \left(\frac{M}{1 - a_1 - a_2}, \frac{M}{1 - a_1 - a_2} \right). \tag{3.6}$$

Referring to the system (3.2), one can define the reversed-time dynamics for $k \le 0$ as follows:

$$\mathbf{x}^*(k-1) = \mathbf{A}^{-1}\mathbf{x}^*(k) - \mathbf{A}^{-1}\mathbf{B}u(k)$$

The solution of the above system, with the control input set to the saturation level $u(k) \equiv \pm M$, is given by the formula

$$\mathbf{x}^*(k) = \mathbf{A}^{-|k|}\mathbf{x}_0 \mp \sum_{i=1}^{|k|} \mathbf{A}^{-i}\mathbf{B}M. \tag{3.7}$$

Using the above formula, it is easy to verify that, since $\mathbf{A}^{-1}$ is a Schur matrix, for any $\mathbf{x}_0 \in \mathbb{R}^n$ one has

$$\lim_{k \to -\infty} \mathbf{x}^*(k) = \mp[(I - \mathbf{A}^{-1})^{-1} - I]\mathbf{B}M = \pm(I - \mathbf{A})^{-1}\mathbf{B}M.$$

In particular, for $\mathbf{x}_0^{\pm}$ given by (3.6), the following property holds:

$$\lim_{k \to -\infty} \phi(k, \mathbf{x}_0^+, -M) = \mathbf{x}_0^-$$

$$\lim_{k \to -\infty} \phi(k, \mathbf{x}_0^-, M) = \mathbf{x}_0^+.$$

Remark 3.1. *It is worth to note that the above property is still valid for systems having a pair of complex unstable eigenvalues, even if no invariant parallelogram exists in this case.*

Let us denote by $\Gamma_{\pm M}^{\sharp}$ the discrete sets of points

$$\Gamma_{\pm M}^{\sharp} = \{\mathbf{x} \in \mathbb{R}^2 : \mathbf{x} = \phi(k, \mathbf{x}_0^{\pm}, \mp M), k \in \mathbb{Z}^-\}. \tag{3.8}$$

Theorem 3.1. *Suppose that the eigenvalues of the state matrix* $\mathbf{A}$ *are complex numbers with* $|\lambda_1| > 1$ *and* $|\lambda_2| > 1$. *Then we have*

$$\mathscr{B}_M^{\sharp} = Co(\Gamma_M^{\sharp} \cup \Gamma_{-M}^{\sharp}).$$

Proof. The proof well be given for the general case of n-dimensional systems, see Theorem 3.3. □

3.1.1 Systems with Positive Real Eigenvalues

The case of systems having two positive eigenvalues can be treated separately since the description of the null controllable region is very simple.

In order to simplify the notation, we set

$$\phi(-|k|, \mathbf{x}_0^{\pm}, \mp M) := \phi_{\pm}(k);$$

together with the sets (3.8), we consider their interpolations

$$\overline{\overline{\Gamma}}_{\pm M} := \bigcup_{j=1}^{\infty} \{\mathbf{x} \in \mathbb{R}^2 : \mathbf{x} = \theta\phi_{\pm}(j-1) + (1-\theta)\phi_{\pm}(j),\ \theta \in [0,1]\}$$

Let us denote by $\phi_{\pm,1}(k), \phi_{\pm,2}(k)$ the components of $\phi(-|k|, \mathbf{x}_0^{\pm}, \mp M)$:

$$\phi(-|k|, \mathbf{x}_0^{\pm}, \mp M) = (\phi_{\pm,1}(k), \phi_{\pm,2}(k)).$$

It can be easily verified that

$$\phi_{\pm,2}(k) = \phi_{\pm,1}(k-1) \tag{3.9}$$

and

$$\phi_{\pm,1}(k) = \pm \frac{-M\lambda_1^k \lambda_2^k + 2M\sum_{i=0}^{k} \lambda_1^i \lambda_2^{k-i} - 2M\sum_{h=0}^{k-1} \lambda_1^h \lambda_2^{k-1-h}}{(\lambda_1 - 1)(\lambda_2 - 1)\lambda_1^k \lambda_2^k}, \tag{3.10}$$

where we have used the identities $a_1 = -\lambda_1\lambda_2$, $a_2 = \lambda_1 + \lambda_2$. The next theorem gives a characterization of the null controllable region for planar systems with two real positive unstable eigenvalues.

Theorem 3.2. *Suppose that the eigenvalues of the state matrix* $\mathbf{A}$ *are real numbers with* $\lambda_1, \lambda_2 > 1$. *Then we have*

$$\partial \mathscr{B}_M^{\natural} = \overline{\overline{\Gamma}}_M \cup \overline{\overline{\Gamma}}_{-M}.$$

Proof. It is easy to see that $\overline{\overline{\Gamma}}_M \cup \overline{\overline{\Gamma}}_{-M}$ is the boundary of a convex domain. This can be proved observing that the incremental ratio

$$\beta_+(k) := \frac{\phi_{+,2}(k) - \phi_{+,2}(k-1)}{\phi_{+,1}(k) - \phi_{+,1}(k-1)}$$

is nonnegative and nondecreasing for any $k \geq 1$. Using the formulas (3.9), (3.10) one can computes $\beta(k)$ explicitly as follows

$$\beta_+(k) = \frac{\lambda_1^k \lambda_2 - \lambda_2^k \lambda_1}{\lambda_1^k - \lambda_2^k} \geq 0;$$

moreover, since $\lambda_1, \lambda_2 > 1$ by assumption, we have

$$\frac{d\beta_+(k)}{dk} = \frac{\lambda_1^k \lambda_2^k (\lambda_1 - \lambda_2)(\log \lambda_1 - \log \lambda_2)}{(\lambda_1^k - \lambda_2^k)^2} \geq 0.$$

By analogous computations, we obtain that the incremental ratio

$$\beta_-(k) := \frac{\phi_{-,2}(k) - \phi_{-,2}(k-1)}{\phi_{-,1}(k) - \phi_{-,1}(k-1)}$$

is nonpositive and nonincreasing. The above analysis shows that the domain enclosed by the polygonal curves $\overline{\overline{\Gamma}}_M$, $\overline{\overline{\Gamma}}_{-M}$ is convex and let us denote this set by $\mathscr{D}$, i.e. $\partial \mathscr{D} = \overline{\overline{\Gamma}}_M \cup \overline{\overline{\Gamma}}_{-M}$. Let be $\mathbf{x} = (x_1, x_2) \in \mathbb{R}^2$ with $x_1 \geq x_2$ and $\mathbf{x} \notin \mathscr{D}$. Since the lower part of the boundary of $\mathscr{D}$, given by $\overline{\overline{\Gamma}}_M$, corresponds to the trajectory of the system driven by the extremal control $u \equiv M$, we can conclude that

$$\phi(k,\mathbf{x},u) \notin \mathscr{D} \; \forall k \in \mathbb{N}, \; \forall u \leq M;$$

in the same way, if $\mathbf{x} = (x_1,x_2) \notin \mathscr{D}$ with $x_1 \leq x_2$, we have

$$\phi(k,\mathbf{x},u) \notin \mathscr{D} \; \forall k \in \mathbb{N}, \; \forall u \geq -M,$$

this meaning that

$$\mathscr{B}_M^\sharp \subseteq \mathscr{D}.$$

To prove the converse inclusion $\mathscr{D} \subseteq \mathscr{B}_M^\sharp$ one can observe that the inner part of $\mathscr{D}$ in covered by a family of omothetic polygonal curves $\overline{\overline{\Gamma}}_{M^*}$, $\overline{\overline{\Gamma}}_{-M^*}$ with $M^* \in (0,M)$. Now if $\mathbf{x} \in \mathscr{D}$, there exists $M_1 < M$ such that $\mathbf{x} \in \overline{\overline{\Gamma}}_{M_1} \cup \overline{\overline{\Gamma}}_{-M_1}$; fixing $u(0) = M^*$ with $M_1 < M^* < M$ one has

$$\phi(1,\mathbf{x},M^*) \in \overline{\overline{\Gamma}}_{M_2} \cup \overline{\overline{\Gamma}}_{-M_2}, \; M_2 < M_1.$$

Iterating the procedure, we can define a decreasing sequence M_k with $\lim_{k\to\infty} M_k = 0$ and an admissible control $u(k) \in \mathscr{W}_M$ such that

$$\phi(k,\mathbf{x},u(k)) \in \overline{\overline{\Gamma}}_{M_{k+1}} \cup \overline{\overline{\Gamma}}_{-M_{k+1}}$$

and as a consequence

$$\lim_{k\to\infty} ||\phi(k,\mathbf{x},u(k))|| = 0. \qquad \square$$

3.2 The SISO n-Dimensional Case

Let us consider a general controllable single-input discrete-time linear system described by

$$\begin{cases} \mathbf{x}(k+1) = \mathbf{A}\mathbf{x}(k) + \mathbf{B}u(k) \\ \mathbf{x}(0) = \mathbf{x}_0 \end{cases} \tag{3.11}$$

with $\mathbf{x} = (x_1,...,x_n) \in \mathbb{R}^n$. In view of the controllability assumption, without loss of generality, the pair $(\mathbf{A},\mathbf{B})$ is given in canonical form. The elements $a_1,...,a_n$ on the last row of the matrix $\mathbf{A}$ are the coefficients of the characteristic polynomial, i.e. $p_{\mathbf{A}}(\lambda) = \sum_{i=1}^n a_i \lambda^{i-1} - \lambda^n$.

For an arbitrary control input $u \in \ell^\infty(\mathbb{Z})$ and a fixed initial condition $\mathbf{x}_0$, the solution of the system will be denoted by $\phi(k,\mathbf{x}_0,u)$. As in the planar case, the input $v(k)$ is assumed to be preceded by a saturating device:

$$u(k) = \mathrm{sat}_M(v(k)), \quad M > 0.$$

Let us denote by $\lambda_1,...,\lambda_n \in \mathbb{R}$ the eigenvalues (not necessarily distinct) of the matrix $\mathbf{A}$. Suppose the system is antistable, that is $|\lambda_i| > 1$ for any $i \in \{1,..,n\}$. Consider the family of surfaces

$$\psi_{\pm M}(k) = x_1(k) + \sum_{i=1}^{n-1} m_i x_{i+1}(k) + q_{\pm M} = 0,\ m_i, q_{\pm M} \in \mathbb{R}$$

and look for all the possible choices of coefficients m_i in order to have

$$\psi_{\pm M}(k) = 0 \Rightarrow \psi_{\pm M}(k+1) = 0 \text{ for } u \equiv \pm M. \tag{3.12}$$

The forward step $\psi_{\pm M}(k+1)$ is given by

$$\begin{aligned}\psi_{\pm M}(k+1) = x_2(k) + \sum_{i=1}^{n-2} m_i x_{i+2}(k) + \\ + m_{n-1}\left(\pm M + \sum_{i=1}^{n} a_i x_i(k)\right) + q_{\pm M}.\end{aligned}$$

Using the equality $\psi_{\pm M}(k) = 0$ and imposing $\psi_{\pm M}(k+1) = 0$, the following set of algebraic equations is obtained

$$\begin{cases} q_{\pm M} = \dfrac{\pm M m_{n-1}}{a_1 m_{n-1} - 1}; \\ 1 + m_{n-1}(-m_1 a_1 + a_2) = 0; \\ m_j + m_{n-1}(-a_1 m_{j+1} + a_{j+2}) = 0, \quad j = 1,...,n-2. \end{cases}$$

Thanks to a recursion property, the system can be rewritten as

$$\begin{cases} m_{n-2} = a_1 m_{n-1}^2 - a_n m_{n-1}, \\ m_{n-3} = a_1^2 m_{n-1}^3 - a_1 a_n m_{n-1}^2 - a_{n-1} m_{n-1}, \\ \vdots \\ m_2 = a_1^{n-3} m_{n-1}^{n-2} - \sum_{j=0}^{n-4} a_1^{n-4-j} a_{n-j} m_{n-1}^{n-3-j}, \\ m_1 = (1 + m_{n-1} a_2)/m_{n-1} a_1, \end{cases}$$

where m_{n-1} satisfies the equation $p_A^*(m_{n-1}) = 0$, with

$$p_A^*(\lambda) = -a_1^{n-1}\lambda^n + 1 + \sum_{j=0}^{n-2} a_1^j a_{j+2} \lambda^{j+1} = 0. \tag{3.13}$$

Proposition 3.2. *Let* ν *be a solution of the characteristic equation* $p_A(\lambda) = \lambda^n - \sum_{j=1}^{n} a_j \lambda^{j-1} = 0$. *Then* ν/a_1 *satisfies*

$$p_A^*(\nu/a_1) = 0.$$

Proof. The statement follows observing that

$$\begin{aligned} p_A^*(\nu/a_1) &= -\frac{\nu^n}{a_1} + 1 + \sum_{j=0}^{n-2} a_1^j a_{j+2} \frac{\nu^{j+1}}{a_1^{j+1}} = \\ &= -\frac{1}{a_1}\left(\nu^n - \sum_{j=1}^{n} a_j \nu^{j-1}\right) = 0. \end{aligned}$$

□

As a consequence, the number of real solutions of Equation (3.13) is exactly the number of real eigenvalues the state matrix $\mathbf{A}$; assuming that $\lambda_j \in \mathbb{R}$ for $j = 1, \ldots, \nu \leq n$, then the real solutions of (3.13) are given by

$$\hat{\lambda}_j = (-1)^{n+1} \frac{\lambda_j}{\Pi_{i=1}^{n} \lambda_i}, \quad j = 1, \ldots, \nu.$$

The above solutions lead to a set of n pairs of parallel hyperplanes; note that for $m_{n-1} = \hat{\lambda}_j$ with $j = 1, \ldots, \nu$, the equation for $q_{\pm M}$ is well-posed. Taking all possible intersections, a convex polytopic set containing the origin as interior point is found. Denote this set as F_M; let us point out that, if $\nu < n$, then F_M is an unbounded set. Due to the invariance property (3.12), if a trajectory starts from a point outside the set F_M, there is no admissible control that can drive it inside.

Proposition 3.3. *Suppose that the system eigenvalues verify* $\lambda_i \in \mathbb{R}$ *and* $|\lambda_i| > 1$ *for any* $i = 1, \ldots, n$. *Then*

$$\mathscr{B}_M^{\sharp} \subseteq F_M.$$

The set ∂F_M intersects the main bisector γ (given by $x_1 = x_2 = \cdots = x_n$) in a pair of symmetric points given by

$$\mathbf{x}_0^{\pm} = \pm \frac{M}{1 - \sum_{i=1}^{n} a_i} \sum_{i=1}^{n} \mathbf{e}_i = \pm M(I - \mathbf{A})^{-1}\mathbf{B}, \tag{3.14}$$

where the sum represents a linear combination of the euclidean basis vectors $\mathbf{e}_i$.

The solution of the reversed-time system ($k \leq 0$)

$$\mathbf{x}^*(k-1) = \mathbf{A}^{-1}\mathbf{x}^*(k) - \mathbf{A}^{-1}\mathbf{B}u(k)$$

with the control input set to the saturation level $u(k) \equiv \pm M$ verifies the identity

$$\mathbf{x}^*(k) = \mathbf{A}^{-|k|}\mathbf{x}_0 \mp \sum_{i=1}^{|k|} \mathbf{A}^{-i}\mathbf{B}M. \tag{3.15}$$

Using the above formula, it is easy to verify that, since $\mathbf{A}^{-1}$ is a Schur matrix, for any $\mathbf{x}_0 \in \mathbb{R}^n$ one has

$$\lim_{k\to-\infty} \mathbf{x}^*(k) = \mp[(I-\mathbf{A}^{-1})^{-1} - I]\mathbf{B}M = \pm(I-\mathbf{A})^{-1}\mathbf{B}M.$$

In particular, for $\mathbf{x}_0^{\pm}$ given by (3.14), the following property holds:

$$\lim_{k\to-\infty} \phi(k,\mathbf{x}_0^+,-M) = \mathbf{x}_0^-$$

$$\lim_{k\to-\infty} \phi(k,\mathbf{x}_0^-,M) = \mathbf{x}_0^+.$$

Remark 3.2. *The above property is valid for any antistable discrete-time system; in particular the points $\mathbf{x}_0^{\pm}$ defined in (3.14) turn out to be the end-points of any reversed-time trajectory corresponding to $u \equiv \pm M$ also for systems having only complex eigenvalues.*

The points $\mathbf{x}_0^{\pm}$ play a central role in the description of the null controllable region $\mathscr{B}_M^{\sharp}$. A unified approach for systems having both real and complex eigenvalues is presented.

Proposition 3.4. *Let $\mathbf{x}_0^{\pm}$ given by (3.14) and consider the countable set of points $\{\gamma(k)\}_{k\le 0} = \left\{\phi(k,\mathbf{x}_0^{\pm},\pm M)\right\}_{k\le 0}$. For $s \in \mathbb{N}$ with $s < n$ there is no s-dimensional affine subspace containing γ.*

Proof. Let us define the vectors $\mathbf{w}_i = \gamma(-i) - \mathbf{x}_0^{\pm}$ and consider the matrix $\mathbf{H}(i_1,...,i_n) \in \mathbb{R}^{n\times n}$, for $i_n > \cdots > i_2 > i_1 \ge 1$, given by

$$\mathbf{H}(i_1,...,i_n) = \begin{bmatrix}\mathbf{w}_{i_1} & \mathbf{w}_{i_2} & \cdots & \mathbf{w}_{i_n}\end{bmatrix}.$$

The statement of the proposition is equivalent to the existence of a multiindex $(i_1^*,\cdots,i_n^*)$ such that

$$\det \mathbf{H}(i_1^*,\cdots,i_n^*) \ne 0.$$

Let us recall that the inverse matrix $\mathbf{A}^{-1}$ has the following structure

$$\mathbf{A}^{-1} = \begin{pmatrix} -\dfrac{\bar{\mathbf{a}}}{a_1} & \dfrac{1}{a_1} \\ \mathbf{I}_{n-1} & \mathbf{0} \end{pmatrix} \qquad \text{with } \bar{\mathbf{a}} = [a_2 \quad a_3 \quad \cdots \quad a_n];$$

using the explicit expression of $\mathbf{A}^{-1}$ in the recursion formula (3.15), the quantities $\mathbf{w}_i$, $i = 1,...,n$, can be easily obtained:

$$\mathbf{w}_1 = \begin{bmatrix} \frac{2M}{a_1} & 0 & 0 & 0 & \cdots & 0 \end{bmatrix}$$

$$\mathbf{w}_2 = \begin{bmatrix} \frac{2M(a_1-a_2)}{a_1^2} & \frac{2M}{a_1} & 0 & 0 & \cdots & 0 \end{bmatrix}$$

$$\mathbf{w}_3 = \begin{bmatrix} \frac{2M\left(a_1^2+a_2^2-a_1(a_2+a_3)\right)}{a_1^3} & \frac{2M(a_1-a_2)}{a_1^2} & \frac{2M}{a_1} & 0 & \cdots & 0 \end{bmatrix}$$

$$\vdots$$

$$\mathbf{w}_{n-1} = \left[\frac{2Mq_{n-1}(a_1,\dots,a_{n-1})}{a_1^{n-1}} \quad \frac{2Mq_{n-2}(a_1,\dots,a_{n-2})}{a_1^{n-2}} \quad \cdots \right.$$

$$\left. \cdots \quad \frac{2M(a_1-a_2)}{a_1^2} \quad \frac{2M}{a_1} \quad 0 \right]$$

$$\mathbf{w}_n = \left[\frac{2Mq_n(a_1,\dots,a_n)}{a_1^n} \quad \frac{2Mq_{n-1}(a_1,\dots,a_{n-1})}{a_1^{n-1}} \quad \cdots \right.$$

$$\left. \cdots \quad \frac{2M\left(a_1^2+a_2^2-a_1(a_2+a_3)\right)}{a_1^3} \quad \frac{2M(a_1-a_2)}{a_1^2} \quad \frac{2M}{a_1} \right]$$

where the function $q_i(\cdot)$ is a suitable homogeneous polynomial of degree $i-1$.

The conclusion follows observing that

$$\det \mathbf{H}(1,\dots,n) = \left(\frac{2M}{a_1}\right)^n \neq 0. \qquad \square$$

The following result can be proved:

Proposition 3.5. *Let* $\mathbf{x}_0 \in \mathbb{R}^n$ *and* $\mathbf{x}_{-1} = \phi(-1,\mathbf{x}_0,u_0)$ *for some* $u_0 \in [-M,M]$. *There exists* $\theta \in (0,1)$ *such that*

$$\mathbf{x}_{-1} = \theta\phi(-1,\mathbf{x}_0,M) + (1-\theta)\phi(-1,\mathbf{x}_0,-M).$$

Proof. The proof is trivial: the coefficient θ is given by

$$\theta = \frac{u_0+M}{2M}. \qquad \square$$

Let us introduce the set-valued function $\Phi_{\pm M}$ defined on $\mathbb{R}^n$ as follows:

$$\Phi_{\pm M}(\mathbf{x}) = \{\mathbf{y} \in \mathbb{R}^n : \exists k \in (-\infty, 0] : \mathbf{y} = \phi(k,\mathbf{x},\pm M)\}.$$

Proposition 3.6. *Given the set* $\Gamma = \Phi_{-M}(\mathbf{x}_0^-) \cup \Phi_M(\mathbf{x}_0^+)$ *together with its convex hull* $S_0 = \mathrm{Co}(\Gamma)$, *there exists* $\delta > 0$ *with* $B_\delta(0) \subset S_0$*; in particular* $|S_0| > 0$.

Proof. The interior part of $\mathrm{Co}(\Gamma)$ is nontrivial since the affine hull of Γ (i.e. the smallest affine space containing the set Γ) is the whole space $\mathbb{R}^n$ by Proposition 3.4; this proves that $|S_0| > 0$. Let $\bar{\mathbf{x}}$ be an interior point of $\mathrm{Co}(\Gamma)$; by symmetry it follows that $-\bar{\mathbf{x}}$ is still an interior point as well as any point $\bar{\mathbf{x}}_\theta$ obtained as convex combination (see Appendix A.1)

$$\bar{\mathbf{x}}_\theta = \theta\bar{\mathbf{x}} - (1-\theta)\bar{\mathbf{x}}, \quad \theta \in [0,1]. \tag{3.16}$$

The result follows observing that the origin verifies (3.16) with $\theta = \dfrac{1}{2}$. □

Theorem 3.3. *Define a monotone recursive sequence of sets* $\{S_j\}_{j\in\mathbb{N}}$

$$S_{j+1} = \mathrm{Co}(\Phi_{-M}(\partial S_j) \cup \Phi_M(\partial S_j)), \tag{3.17}$$

where S_0 *is given by* $S_0 = \mathrm{Co}(\Gamma)$. *The null controllable region of the system is given by*

$$\mathscr{B}_M^\sharp = \lim_{j\to\infty} S_j = \bigcup_{j\in\mathbb{N}} S_j =: S_\infty$$

Proof. First prove that $\mathscr{B}_M \subseteq S_\infty$. Suppose $\mathbf{x}_0 \in \mathscr{B}_M$ and $\mathbf{x}_0 \notin S_0$. Since $\mathbf{x}_0 \in \mathscr{B}_M$, there exists $u \in \mathscr{U}_M$ such that

$$\lim_{k\to\infty} \|\phi(k, \mathbf{x}_0, u)\| = 0.$$

In particular there exists $\bar{k}$ such that $\bar{\mathbf{x}}_{\bar{k}} := \phi(\bar{k}, \mathbf{x}_0, u) \notin S_0$ and $\bar{\mathbf{x}}_k := \phi(k, \mathbf{x}_0, u) \in S_0$ for any $k \geq \bar{k}+1$. Now

$$\bar{\mathbf{x}}_{\bar{k}} = \phi(-1, \bar{\mathbf{x}}_{\bar{k}+1}, u).$$

By Proposition 3.5, there exists $0 \leq \theta \leq 1$ such that

$$\bar{\mathbf{x}}_{\bar{k}} = \theta\phi(-1, \mathbf{x}_{\bar{k}+1}, M) + (1-\theta)\phi(-1, \bar{\mathbf{x}}_{\bar{k}+1}, -M). \tag{3.18}$$

Since $\bar{\mathbf{x}}_{\bar{k}+1} \in S_0$ and by construction $\phi(-1, \bar{\mathbf{x}}_{\bar{k}+1}, M) \in \mathrm{Co}(\Phi_{-M}(\partial S_0) \cup \Phi_M(\partial S_0))$, identity (3.18) implies that

$$\bar{\mathbf{x}}_{\bar{k}} \in S_1.$$

Iterating the above procedure it can be easily proved that

$$\mathbf{x}_0 \in S_{\bar{k}} \subset S_\infty.$$

Let us show that the converse inclusion $S_\infty \subseteq \mathscr{B}_M^\sharp$ holds too. Saying that $\mathbf{x}_0 \in S_\infty$ means that exists k such that $\mathbf{x}_0 \in S_k$. Moreover, by definition, there exists a control function in $\mathscr{W}_M^*$ having at most k switches such that the solution driven by it enters S_0 in finite time. It remains only to prove that $S_0 \subseteq \mathscr{B}_M^\sharp$. Let $\mathbf{x}_0 \in S_0$. Since the solutions are continuously dependant on the system parameters, the interior part of S_0 is completely covered by a family of homothetic closed surfaces which are generated starting from the points $\mathbf{x}_{0,M^*} = \pm(I - \mathbf{A})^{-1}\mathbf{B}M^*$ and switching the control between the values M^* as M^* varies in $[0,M]$. Denote these sets as Z_{M^*}. In particular it holds

$$Z_{M^*} = \partial S_{0,M^*},$$

where

$$S_{0,M^*} = \mathrm{Co}(\Phi_{-M^*}(\mathbf{x}^-_{0,M^*}) \cup \Phi_{M^*}(\mathbf{x}^+_{0,M^*}))$$

and

$$S_0 = \bigcup_{0 \le M^* \le M} Z_{M^*}, \quad Z_{M_1} \cap Z_{M_2} = \emptyset \quad \text{for } M_1 \neq M_2.$$

Let $\mathbf{x}_0^* \in Z_{M^*}$; by the Carathéodory theorem (see Appendix A.1), there exist $\theta_j > 0$, $1 < j < N+1$, with $\sum_{j=1}^{N+1} \theta_j = 1$ and $k_j \in (-\infty, 0]$ such that

$$\mathbf{x}_0^* = \sum_{j=1}^{j^*} \theta_j \phi(k_j, \mathbf{x}^+_{0,M^*}, M^*) + \\ + \sum_{j=j^*+1}^{N+1} \theta_j \phi(k_j, \mathbf{x}^-_{0,M^*}, -M^*), \tag{3.19}$$

with $j^* \in [0, N+1]$. Note that the elements of S_∞ obtained as $\lim_{k \to -\infty} \phi(k, \mathbf{x}, M)$, for some $\mathbf{x}$, belong to ∂S_∞. It follows from (3.19) that

$$\mathbf{x}^*(t) = \sum_{j=1}^{j^*} \theta_j \phi(k + k_j, \mathbf{x}^+_{0,M^*}, M^*) + \\ + \sum_{j=j^*+1}^{N+1} \theta_j \phi(k + k_j, \mathbf{x}^-_{0,M^*}, -M^*)$$

satisfies the system equations for initial datum $\mathbf{x}_0^*$ and control input $u^*(k) = M^*$ $(\sum_{j=1}^{j^*} \theta_j - \sum_{j=j^*+1}^{N+1} \theta_j)$. This means that any point in Z_{M^*} lies on a trajectory associated to a suitable admissible control input. As a consequence, from the invariance of the closed curve $\Phi_{-M^*}(\mathbf{x}^-_{0,M^*}) \cup \Phi_{M^*}(\mathbf{x}^+_{0,M^*})$, the invariance of the whole set Z_{M^*} can be deduced. Now, if $\mathbf{x}_0 \in S_0$, there exist $M_1^* < M$ such that $\mathbf{x}_0$ lies on the invariant surface $Z_{M_1^*}$ under a suitable control input u^* with $|u^*| \le M_1^* < M$. In order to stabilize the system, it suffices to make the trajectory jump to a lower level $Z_{M_2^*}$ with $M_2^* < M_1^*$ increasing the control norm. This concludes the proof. □

3.3 The MIMO Case

As for continuous-time systems, the description of the null controllable for discrete-time MIMO systems can be obtained with a direct method as well as by a method based on lower-order single input subsystems. Since the technical results are analogous to those presented in Section 2.4, some details have been omitted here. Let us consider the following multi input linear system

$$\begin{cases} \mathbf{x}(k+1) = \mathbf{A}\mathbf{x}(k) + \mathbf{B}\mathbf{u}(k) \\ \mathbf{x}(0) = \mathbf{x}_0 \end{cases} \tag{3.20}$$

with $m < n$, $\mathbf{A} \in \mathbb{R}^{n \times n}$, $\mathbf{B} \in \mathbb{R}^{n \times m}$, $\mathbf{x} \in \mathbb{R}^n$ and $\mathbf{u} \in \mathbb{R}^m$; the plant is assumed to be controllable, i.e.

$$\mathrm{rank}[\mathbf{B}\ \mathbf{A}\mathbf{B}\ \mathbf{A}^2\mathbf{B} \cdots \mathbf{A}^{n-1}\mathbf{B}] = n \tag{3.21}$$

and antistable, that is $|\lambda_i| > 1$ for any eigenvalue λ_i, $i = 1, ..., n$, of the matrix $\mathbf{A}$. We will use the following notation:

$$\mathbf{B} = [\mathbf{B}_1\ \mathbf{B}_2\ \cdots\ \mathbf{B}_m], \quad \mathbf{u} = \begin{bmatrix} u_1 \\ u_2 \\ \vdots \\ u_m \end{bmatrix},$$

where $\mathbf{B}_i \in \mathbb{R}^n$ and $u_i \in \mathbb{R}$ for any $i = 1, ..., m$.

3.3.1 Method of Lower-Order Single Input Subsystems

According to (3.20), we can define a family of single input systems Σ_i, $i = 1, .., m$, as follows

$$\Sigma_i = \begin{cases} \mathbf{x}(k+1) = \mathbf{A}\mathbf{x}(k) + \mathbf{B}_i u_i(k) \\ \mathbf{x}(0) = \mathbf{x}_0 \end{cases}$$

Let $n_i \in \mathbb{N}$ be the dimension of the controllability subspace for the system Σ_i,

$$n_i := \mathrm{rank}[\mathbf{B}_i\ \mathbf{A}\mathbf{B}_i\ \mathbf{A}^2\mathbf{B}_i\ \cdots\ \mathbf{A}^{n-1}\mathbf{B}_i] \le n.$$

The system Σ_i can be transformed into a controllable/uncontrollable subsystems decomposition by a linear coordinates transformation and one can easily compute the maximal stability region $\mathscr{D}_M^{(i),\sharp}$ of the single input system Σ_i (see Section 2.3 for details). By construction, the set $\mathscr{D}_M^{(i),\sharp}$ is a convex region contained in a linear subspace $V_i \subset \mathbb{R}^n$ with $\dim(V_i) = n_i$.

Theorem 3.4. *The null controllable region $\mathscr{B}_M^\sharp$ for the multi input system (3.20) is given by*

$$\mathscr{B}_M^\sharp = \sum_{i=1}^m \mathscr{D}_M^{(i),\sharp} = \left\{ \mathbf{x} \in \mathbb{R}^n : \mathbf{x} = \mathbf{x}_1 + \mathbf{x}_2 + \cdots + \mathbf{x}_m,\ \mathbf{x}_i \in \mathscr{D}_M^{(i),\sharp}\ \forall\, i = 1, ..., m \right\}.$$

Proof. The set $\sum_{i=1}^{m} \mathscr{D}_M^{(i),\sharp}$, as it is defined as a finite sum of convex sets, is still convex (see for instance [33]). Let $\mathbf{x} \in \sum_{i=1}^{m} \mathscr{D}_M^{(i),\sharp}$, with $\mathbf{x} = \sum_{i=1}^{m} \mathbf{x}_i$. Since $\mathbf{x}_i \in \mathscr{D}_M^{(i),\sharp}$, by construction, there exists $u_i \in \mathscr{W}_M$ such that, defining $\phi_i(k, \mathbf{x}_0, u(k))$ the dynamics of the system Σ_i driven by the control u, we have

$$\lim_{k \to \infty} ||\phi_i(k, \mathbf{x}_i, u_i(k)|| = 0.$$

Now, since

$$\phi(k, \mathbf{x}, \mathbf{u}(k) = \sum_{i=1}^{m} \phi_i(k, \mathbf{x}_i, u_i(k)),$$

we can deduce the existence of $\mathbf{u} \in \mathscr{W}_{\mathbf{M}}$ such that

$$\lim_{k \to \infty} ||\phi_i(k, \mathbf{x}, \mathbf{u}(k))|| = 0;$$

in other words we have $\sum_{i=1}^{m} \mathscr{D}_M^{(i),\sharp} \subseteq \mathscr{B}_M^{\sharp}$. Let us show that the converse inclusion $\mathscr{B}_M^{\sharp} \subseteq \sum_{i=1}^{m} \mathscr{D}_M^{(i),\sharp}$ holds too. Suppose that $\mathbf{x} \in \mathscr{B}_M^{\sharp}$; by definition, there exists $u = (u_1, \ldots, u_m)$ with $|u_i| \leq M$ such that

$$\lim_{k \to \infty} ||\phi_i(k, \mathbf{x}, \mathbf{u}(k))|| = 0. \tag{3.22}$$

Using the explicit formula for the solution we have

$$\phi_i(k, \mathbf{x}, \mathbf{u}(k)) = \mathbf{A}^k \mathbf{x} + \sum_{j=0}^{k-1} \mathbf{A}^{k-1-j} \sum_{i=1}^{m} \mathbf{B}_i u_i(j).$$

From (3.22) we obtain

$$\lim_{k \to \infty} \mathbf{A}^k \left(\mathbf{x} + \sum_{j=0}^{k-1} \mathbf{A}^{-1-j} \sum_{i=1}^{m} \mathbf{B}_i u_i(j) \right) = 0$$

and necessarily one has

$$\lim_{k \to \infty} \sum_{j=0}^{k-1} \mathbf{A}^{-1-j} \sum_{i=1}^{m} \mathbf{B}_i u_i(j) = -\mathbf{x}.$$

Now, since $||\sum_{i=1}^{m} \mathbf{B}_i u_i(j)|| \leq M$, $\sum_{i=1}^{m} ||\mathbf{B}_i|| < \infty$ and $\sum_{j=0}^{\infty} ||\mathbf{A}^{-1-j}|| < \infty$, the following limits exist $\forall i = 1, \ldots, m$ and they are finite:

$$\lim_{k \to \infty} \sum_{j=0}^{k-1} \mathbf{A}^{-1-j} \mathbf{B}_i u_i(j) = -\mathbf{x}_i.$$

As a consequence one has

$$\lim_{k\to\infty} \mathbf{A}^k \left(\mathbf{x}_i + \sum_{j=0}^{k-1} \mathbf{A}^{-1-j} \mathbf{B}_i u_i(j) \right) = 0 \ \forall i = 1,...,m,$$

that is $\mathbf{x}_i \in \mathscr{D}_M^{(i),\sharp}$; now, since by construction $\mathbf{x} = \sum_{i=1}^m \mathbf{x}_i$, we have $\mathscr{B}_M^\sharp \subseteq \sum_{i=1}^m \mathscr{D}_M^{(i),\sharp}$ and the statement is proved. □

3.3.2 Direct Method

The direct method for the description of the multi input null controllable region follows the same steps presented in the single input case. In particular, the null controllable region can be computed with an iterative procedure combining reversed-time evolution with extremal controls and set convexification.

Let us denote by $\mathscr{W}_M^{**}$ the set of extremal controls; to this purpose we recall that a control input $\mathbf{u} = (u_1,...,u_m) \in \mathscr{W}_M$ is extremal if $u_i \equiv \pm M \ \forall i = 1,...,m$. Using the explicit formula for the solution $\phi(k,\mathbf{x},\mathbf{u}(k))$ one can verify that $\forall \mathbf{u} \in \mathscr{W}_M^{**}$

$$\lim_{k\to-\infty} \phi(k,\mathbf{x},\mathbf{u}^*) = (\mathbf{I}-\mathbf{A})^{-1}\mathbf{B}\mathbf{u}^* \ \forall \mathbf{x} \in \mathbb{R}^n.$$

Fixing $u_1^* \equiv M$, there are 2^{m-1} possible combinations for the other components of an extremal control $\mathbf{u} \in \mathscr{W}_M^{**}$; in this way, we can define a set of 2^{m-1} points

$$\mathbf{x}_{M,\mathbf{u}} = (\mathbf{I}-\mathbf{A})^{-1} \left(\mathbf{B}_1 + \sum_{i=2}^m \pm \mathbf{B}_i \right) M$$

such that

$$\lim_{k\to-\infty} \phi(k, \mp\mathbf{x}_{M,\mathbf{u}}, \pm\mathbf{u}) = \pm\mathbf{x}_{M,\mathbf{u}} \ \forall \mathbf{u} \in \mathscr{W}_{M,+}^{**}$$

where it has been set $\mathscr{W}_{M,+}^{**} := \{\mathbf{u} \in \mathscr{W}_M^{**} : u_1 = M\}$. For $\mathbf{u} \in \mathscr{W}_{M,+}^{**}$ we define the family of polygonal curves having vertices

$$\gamma_{M,\mathbf{u}}(k) := \phi(k, -\mathbf{x}_{M,\mathbf{u}}, \mathbf{u}), \ k \in \mathbb{Z} \setminus \mathbb{N}$$

and the corresponding supports

$$\Gamma_{M,\mathbf{u}} := \{\mathbf{x} \in \mathbb{R}^n : \mathbf{x} = \phi(k, -\mathbf{x}_{M,\mathbf{u}}, \mathbf{u}), \ k \in \mathbb{Z} \setminus \mathbb{N}\}. \tag{3.23}$$

Proposition 3.1. *There exists at least one extremal control $\bar{\mathbf{u}} \in \mathscr{W}_{M,+}^{**}$ such that, for $s < n$ there is no s-dimensional affine subspace containing $\gamma_{M,\bar{\mathbf{u}}}(k)$.*

Proof. For any fixed $\mathbf{u} \in \mathscr{W}_{M,+}^{**}$ the multi input system $(\mathbf{A},\mathbf{B})$ can be regarded as the single input system $(\mathbf{A},\mathbf{B}_\mathbf{u})$ driven by $u \equiv M$ with

$$\mathbf{B}_{\mathbf{u}} = \mathbf{B}_1 + \sum_{i=1}^{m} \pm \mathbf{B}_i; \tag{3.24}$$

the signs in the above formula are fixed according to the signs of the components u_i, $i = 2,...m$. Since the plant $(\mathbf{A},\mathbf{B})$ is assumed to be controllable, there exists necessarily $\bar{\mathbf{u}} \in \mathscr{U}^{**}_{M,+}$ such that the single input system $(\mathbf{A},\mathbf{B}_{\bar{\mathbf{u}}})$, with $\mathbf{B}_{\bar{\mathbf{u}}}$ defined by (3.24), is controllable as well. Now the curve $\gamma_{M,\bar{\mathbf{u}}}(t)$ can be regarded as the reversed-time evolution of the single input system $(\mathbf{A},\mathbf{B}_{\bar{\mathbf{u}}})$ driven by the control $u \equiv M$ and the conclusion follows applying the results given in Proposition 3.4 to such polygonal curve. □

Introducing the set-valued function $\Phi_{\pm\mathbf{u}}$, $\mathbf{u} \in \mathscr{W}^{**}_{M,+}$, defined on $\mathbb{R}^n$ as follows

$$\Phi_{\mathbf{u}}(\mathbf{x}) = \{\mathbf{y} \in \mathbb{R}^n : \exists k \in \mathbb{Z}\setminus\mathbb{N} : \mathbf{y} = \phi(k,\mathbf{x},\mathbf{u})\},$$

one can design a monotone recursive sequence of sets $\{S_j\}_{j\in\mathbb{N}}$

$$S_{j+1} = \mathrm{Co}\left(\bigcup_{\mathbf{u}\mathscr{W}^{**}_{M,+}} (\Phi_{-\mathbf{u}}(\partial S_j) \cup \Phi_{\mathbf{u}}(\partial S_j))\right), \tag{3.25}$$

where S_0 is given by $S_0 = \mathrm{Co}\left(\bigcup_{\mathbf{u}\mathscr{W}^{**}_{M,+}} -\Gamma_{M,\mathbf{u}} \cup \Gamma_{M,\mathbf{u}}\right)$ with $\Gamma_{M,\mathbf{u}}$ given by (3.23).

Theorem 3.5. *The null controllable region for the multi input system (3.20) is*

$$\mathscr{B}^{\sharp}_M = \lim_{j\to\infty} S_j = \bigcup_{j\in\mathbb{N}} S_j =: S_\infty$$

The proof is omitted. It can be derived using Proposition 3.1 and applying the same arguments used to prove Theorem 3.3.

3.4 Examples

Example 3.1. Let us consider the antistable system described by the matrices

$$\mathbf{A} = \begin{pmatrix} 0 & 1 \\ -3.15 & 3.6 \end{pmatrix}, \qquad \mathbf{B} = \begin{pmatrix} 0 \\ 1 \end{pmatrix}.$$

with $M = 1$. The open-loop eigenvalues are

$$\lambda_1 = 1.5 \qquad \lambda_2 = 2.1$$

and the initial points $\mathbf{x}_0^{\pm} = \pm M(\mathbf{I} - \mathbf{A})^{-1}\mathbf{B}$ are given by

$$\mathbf{x}_0^{\pm} = \pm(1.81, 1.81).$$

The next Fig.3.1 shows the null controllable region for the above system.

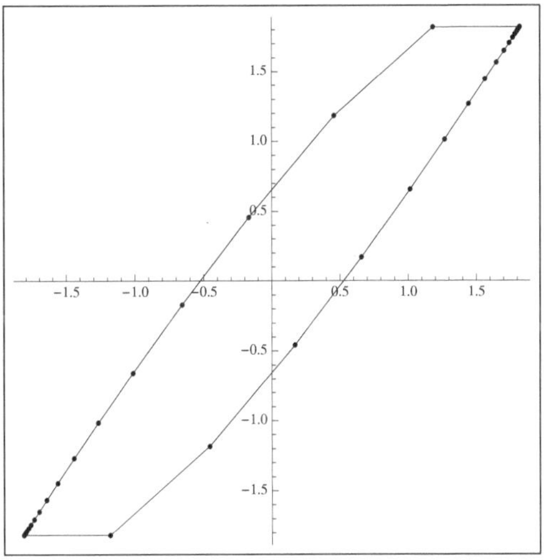

Fig. 3.1. Boundary of the null controllable region for real positive eigenvalues $\lambda_1 = 1.5, \quad \lambda_2 = 2.1$ and saturation level $M = 1$.

Example 3.2. Let us consider the system expressed by the following parameters

$$\mathbf{A} = \begin{pmatrix} 0 & 1 \\ 6.12 & 1.9 \end{pmatrix}, \qquad \mathbf{B} = \begin{pmatrix} 0 \\ 1 \end{pmatrix}.$$

The system eigenvalues are

$$\lambda_1 = -1.7 \qquad \lambda_2 = 3.6$$

and the initial points $\mathbf{x}_0^{\pm} = \pm M(\mathbf{I} - \mathbf{A})^{-1}\mathbf{B}$ are given by

$$\mathbf{x}_0^{\pm} = \mp(1.42, 1.42).$$

A picture of the null controllable region for the system is shown in the Fig. 3.2 below.

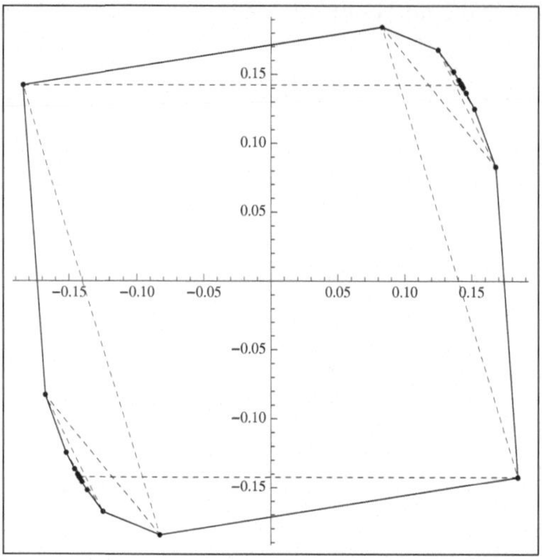

Fig. 3.2. Boundary of the null controllable region for real eigenvalues $\lambda_1 = -1.7, \quad \lambda_2 = 3.6$ and saturation level $M = 1$. Dashed lines represent extremal trajectories corresponding to control inputs $u \equiv \pm M$.

Example 3.3. Let us consider the following system matrices

$$\mathbf{A} = \begin{pmatrix} 0 & 1 \\ -2.02 & 1.8 \end{pmatrix}, \qquad \mathbf{B} = \begin{pmatrix} 0 \\ 1 \end{pmatrix}.$$

The open-loop eigenvalues are a pair of complex conjugate numbers:

$$\lambda_1 = 0.9 + 1.1\mathrm{j} \qquad \lambda_2 = 0.9 - 1.1\mathrm{j};$$

the initial points $\mathbf{x}_0^{\pm} = \pm M(\mathbf{I} - \mathbf{A})^{-1}\mathbf{B}$ are given by

$$\mathbf{x}_0^{\pm} = \pm(0.82, 0.82).$$

A picture of the null controllable region for the system is showed in Fig. 3.3.

Example 3.4. In the last example the system matrices are set as

$$\mathbf{A} = \begin{pmatrix} 0 & 1 \\ -1.94 & -2.6 \end{pmatrix}, \qquad \mathbf{B} = \begin{pmatrix} 0 \\ 1 \end{pmatrix}.$$

The eigenvalues of $\mathbf{A}$ are a pair of conjugate complex numbers given by

$$\lambda_1 = -1.3 + 0.5\mathrm{j} \qquad \lambda_2 = -1.3 - 0.5\mathrm{j}$$

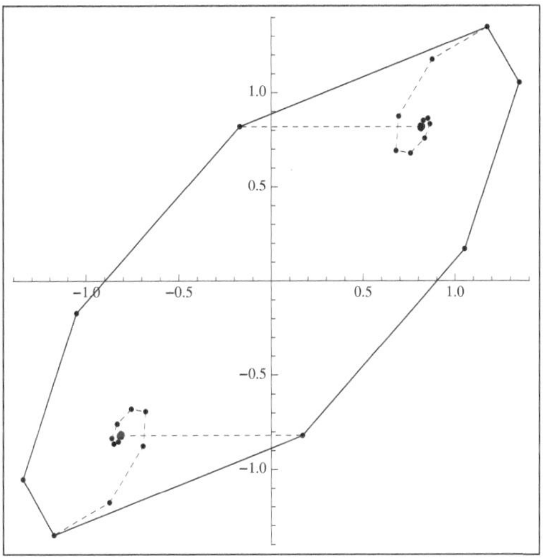

Fig. 3.3. Boundary of the null controllable region for complex eigenvalues $\lambda_1 = 0.9 + 1.1\mathrm{j}$, $\lambda_2 = 0.9 - 1.1\mathrm{j}$ and saturation level $M = 1$. Dashed lines represent extremal trajectories corresponding to control inputs $u \equiv \pm M$.

and the initial points $\mathbf{x}_0^{\pm} = \pm M(\mathbf{I} - \mathbf{A})^{-1}\mathbf{B}$ are given by

$$\mathbf{x}_0^{\pm} = \pm(0.18, 0.18).$$

Figure 3.4 shows a picture of the null controllable region for the system.

Remark 3.3. *In the first example (systems with real positive eigenvalues) the null controllable turns out to be equal to the region bounded by $\Gamma_M \cup \Gamma_{-M}$; on the other hand, such region is in general not convex for systems with real negative or complex eigenvalues and $\mathscr{B}_M$ is equal to the convex hull $Co(\Gamma_M \cup \Gamma_{-M})$ as showed in the other examples.*

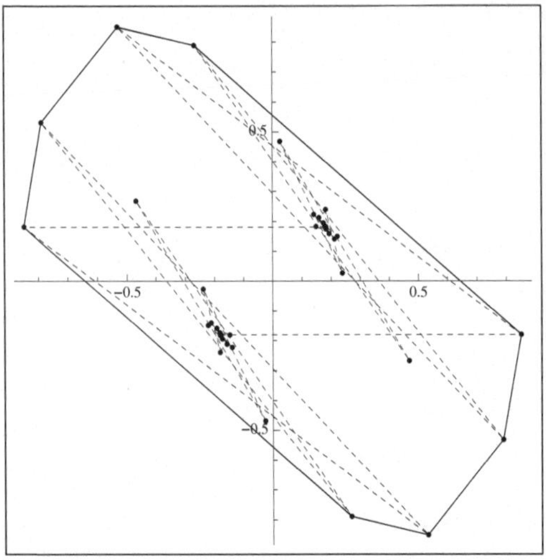

Fig. 3.4. Boundary of the null controllable region for complex eigenvalues $\lambda_1 = -1.3 + 0.5\mathrm{j}$, $\lambda_2 = -1.3 - 0.5\mathrm{j}$ and saturation level $M = 1$. Dashed lines represent extremal trajectories corresponding to control inputs $u \equiv \pm M$.

Part II
Design Issues

Chapter 4
Control Design Issues: Continuous-Time Plants

Earlier in the book, attention has been focussed on the analysis of the basin of attraction of a linear system subject to actuator saturation, and mostly on its geometrical determination and numerical computation. It is now time to drive our attention to the stabilization of a linear plant on its null controllable region with saturating actuators.

As a first look to the problem, a planar unstable plant without uncertainties is considered (the extension to linear plants of arbitrary finite dimensions is discussed next). A design technique is proposed for finding a linear state feedback controller with the property of having nonincreasing norm along the closed-loop system trajectories [36]. Interestingly, the set of initial conditions satisfying the saturation constraint turns out to be invariant for the closed-loop system evolution (results on invariant sets in control theory can be found in [37] and [38]). In particular, the region of attraction associated to such controller is an unbounded strip and it can be straightforwardly characterized. Furthermore, it is shown show how, once the saturation level is fixed to $M > 0$, it is possible to split the controller into a finite number of saturating components. The number of components can be a priori determined for any fixed compact set of initial data. By a practical point of view, this scheme can be regarded as the model of a plant equipped with a controller followed by a finite number of saturating devices with the same structure.

In the presence of disturbances affecting the plant, we are primarily concerned on boundedness of trajectories in face of the perturbing term, and mostly on achieving disturbance rejection, i.e. on designing feedback laws having the ability to completely reject at least a given class of disturbances.

In the wide literature addressing the problem of disturbance rejection for linear systems subject to actuator saturation, an interesting research line considers disturbances that are magnitude bounded. This line complements the research thrust addressing $\mathscr{L}_p$ disturbances [20] [21]. In the former framework, [22] proved that semiglobal practical stabilization for a linear system subject to actuator saturation and input additive disturbances can be achieved as long as the open loop system is not exponentially unstable. For the same class of systems, Lin [7] constructed nonlinear feedback laws that achieve global practical stabilization. Very recently, it has been proved in [23] that a 2-dimensional linear systems subject to actuator

M.L. Corradini et al.: Control Systems with Saturating Inputs, LNCIS 424, pp. 55–94.
springerlink.com

saturation and bounded input additive disturbances can be globally practically stabilized by linear state feedback.

Variable Structure Control techniques have been rarely used to address the problem of robustly controlling plants with saturating actuators. It is worth mentioning that higher order sliding mode techniques have been effectively employed to achieve plant stabilization in the presence of unmodeled actuator dynamics [39]. In this chapter, a design technique will be described ensuring the robust asymptotical stabilization of a Multi-Input linear plant [40] [41]. The time-varying state feedback control law is derived imposing the achievement of a sliding motion onto a suitable time-varying sliding surface [42] [43]. It can be proved that a constructive procedure exists for designing the surface as to guarantee the asymptotical stabilization of the plant in the presence of bounded matched uncertainties, under the usual assumption of the saturation threshold being larger than the bound on uncertainties.

4.1 Invariant Strips and Linear Feedback Laws

4.1.1 Planar Systems

Consider a 2-dimensional controllable SISO linear system having, without loss of generality, the following structure

$$\begin{cases} \dot{\mathbf{x}} = \mathbf{A}\mathbf{x} + \mathbf{b}u \\ \mathbf{x}(0) = \mathbf{x}_0 \end{cases} \tag{4.1}$$

where

$$\mathbf{A} = \begin{pmatrix} 0 & 1 \\ a_1 & a_2 \end{pmatrix}, \qquad \mathbf{b} = \begin{pmatrix} 0 \\ 1 \end{pmatrix},$$

$\mathbf{x}(t) = (x_1(t), x_2(t)) \in \mathrm{C}^0([0,+\infty), \mathbb{R}^2)$ is the state vector and $u(t) \in \mathrm{C}^0_\sharp([0,+\infty), \mathbb{R})$ is the input variable.

We will suppose $a_1 > 0$; this implies that the eigenvalues of $\mathbf{A}$ are real numbers with opposite signs. Input saturation is assumed to be present in the system; in particular any input u is subject to the constraint

$$u = \mathrm{sat}_M(v),\ M > 0, \tag{4.2}$$

where $\mathrm{sat}_M(s)$ is given by (1.3) and $u(t) \in \mathrm{C}^0_\sharp([0,+\infty), \mathbb{R})$.

Lemma 4.1. *Given the system (4.1) with $a_1 > 0$ there exists a linear feedback $u = \langle \mathbf{K}, \mathbf{x} \rangle$ such that the closed-loop system driven by $\mathbf{A} + \mathbf{b}\mathbf{K}^T$ is asymptotically stable in the origin and moreover the following condition holds:*

- *if $\langle \mathbf{K}, \mathbf{x} \rangle \geq 0$ then $\langle \mathbf{K}, \dot{\mathbf{x}} \rangle \leq 0$,*
- *if $\langle \mathbf{K}, \mathbf{x} \rangle \leq 0$ then $\langle \mathbf{K}, \dot{\mathbf{x}} \rangle \geq 0$.*

Proof. Given a saturation level $M > 0$, we consider the level set $\langle \mathbf{K}, \mathbf{x} \rangle = M$; our aim is to construct $\mathbf{K}$ such that $\langle \mathbf{K}, \mathbf{A}\mathbf{x} + \mathbf{b}u \rangle \leq 0$ over the previous set, that is

$$\langle \mathbf{K}, \mathbf{A}\mathbf{x} + M\mathbf{b} \rangle \leq 0 \ \forall \mathbf{x} \text{ such that } \langle \mathbf{K}, \mathbf{x} \rangle = M. \tag{4.3}$$

Let us denote by k_1, k_2 the components of $\mathbf{K}$. In order to achieve condition (4.3), the two affine subspaces $\{\mathbf{x} \in \mathbb{R}^2 : \langle \mathbf{K}, \mathbf{x} \rangle = M\}$ and $\{\mathbf{x} \in \mathbb{R}^2 : \langle \mathbf{K}, \mathbf{A}\mathbf{x} + M\mathbf{b} \rangle = 0\}$ must have no intersection, in particular they must be parallel, and this can hold if and only if $\mathbf{K}$ is an eigenvector of the matrix $\mathbf{A}^T$. Moreover the corresponding eigenvalue λ must be real with $\lambda \leq -k_2$. Parallelism between the spaces $\{\mathbf{x} \in \mathbb{R}^2 : \langle \mathbf{K}, \mathbf{x} \rangle = -M\}$ and $\{\mathbf{x} \in \mathbb{R}^2 : \langle \mathbf{K}, \mathbf{A}\mathbf{x} - M\mathbf{b} \rangle = 0\}$ gives the same condition on $\mathbf{K}$. The eigenvalue λ is characterized by

$$\lambda k_1 = a_1 k_2 \tag{4.4}$$

$$\lambda k_2 = k_1 + a_2 k_2. \tag{4.5}$$

From the first equation we get

$$\lambda = \frac{a_1 k_2}{k_1}$$

and we can transform the last one in a second degree equation for the ratio k_2/k_1

$$a_1 \frac{k_2^2}{k_1^2} - a_2 \frac{k_2}{k_1} - 1 = 0. \tag{4.6}$$

Since $a_1 > 0$, in order to obtain stabilization of the system we must require $k_1 < 0$ (recall that $a_1 + k_1$ represents the opposite of the product between the eigenvalues of the closed-loop system). Moreover we will assume $k_2 < 0$ in both cases $a_2 > 0$ and $a_2 < 0$. The positive solution of (4.6) is given by

$$\frac{k_2}{k_1} = \frac{a_2 + \sqrt{a_2^2 + 4a_1}}{2a_1}. \tag{4.7}$$

4.1.1.1 CASE $\mathbf{tr}(\mathbf{A}) > 0$

If $a_2 > 0$ we search for $\mathbf{K}$ given by

$$k_1 = -a_1 - \eta, \quad k_2 = -a_2 - \varepsilon \quad \text{with } \eta, \varepsilon > 0. \tag{4.8}$$

Using these expressions of the coefficients of $\mathbf{K}$ in (4.7), we obtain

$$\eta = \frac{a_1 a_2 - a_1 \sqrt{a_2^2 + 4a_1} + 2a_1 \varepsilon}{a_2 + \sqrt{a_2^2 + 4a_1}} \tag{4.9}$$

which is positive if we take

$$\varepsilon > \frac{\sqrt{a_2^2+4a_1}-a_2}{2} \ (>0). \tag{4.10}$$

4.1.1.2 CASE tr(A) < 0

In the complementary case $a_2 < 0$ we will choose k_1 as before and

$$k_2 = -\varepsilon < 0. \tag{4.11}$$

We find

$$\eta = \frac{-a_1a_2 - a_1\sqrt{a_2^2+4a_1}+2a_1\varepsilon}{a_2+\sqrt{a_2^2+4a_1}} \tag{4.12}$$

and then ε must satisfy

$$\varepsilon > \frac{\sqrt{a_2^2+4a_1}+a_2}{2} \ (>0). \tag{4.13}$$

This concludes the proof of lemma. □

Remark 4.1. *A vector* $\mathbf{K}$ *with the desired properties can be found also in the border case* $a_1 > 0, a_2 = 0$*; it is enough to choose* $\mathbf{K} = (-a_1-\eta, -\varepsilon)$ *with*

$$\varepsilon, \eta > 0 \text{ and } \frac{\varepsilon^2}{(-a_1-\eta)^2} = \frac{1}{a_1}.$$

Theorem 4.1. *Given the system (4.1) with* $a_1 > 0$ *there exists a vector* $\overline{\mathbf{K}} \in \mathbb{R}^2$ *such that for any* $M > 0$

1. $\Omega_M = \{\mathbf{x} \in \mathbb{R}^2 : |\langle \overline{\mathbf{K}}, \mathbf{x}\rangle| \leq M\}$ *is an invariant set for the solution* $\phi(t, \mathbf{x}_0, \overline{\mathbf{K}})$*;*
2. *if* $\mathbf{x}_0 \in \Omega_M$ *we have* $\lim_{t\to\infty} \|\phi(t, \mathbf{x}_0, \overline{\mathbf{K}})\| = 0$ *under the saturation constraint (2.2).*

The expression of the coefficients of $\overline{\mathbf{K}}$ *is given by (4.8)-(4.9)-(4.10) and (4.11)-(4.12)-(4.13) for* $a_2 > 0$ *and* $a_2 < 0$*, respectively. See Remark 4.1 for the case* $a_2 = 0$.

Proof. We give the proof for $a_2 > 0$. We set $\overline{\mathbf{K}} = \overline{\mathbf{K}}(\varepsilon) = (-a_1 - \eta(\varepsilon), -a_2 - \varepsilon)$ with ε satysfying (4.10). Taking the initial datum $\mathbf{x}_0 \in \Omega_M$, since $\langle \mathbf{K}, \dot{\mathbf{x}}\rangle \leq 0$ on the boundary component $\{x \in \mathbb{R}^2 : \langle \mathbf{K}, \mathbf{x}\rangle = M\}$, the control cannot exceed the level M; similarly $\langle \mathbf{K}, \dot{\mathbf{x}}\rangle \geq 0$ on $\{x \in \mathbb{R}^2 : \langle \mathbf{K}, \mathbf{x}\rangle = -M\}$ and so Ω_M is an invariant set.

Moreover the closed-loop system $\mathbf{A} + \mathbf{b}\overline{\mathbf{K}}^T$ has eigenvalues with negative real parts by construction, so that asymptotic stability is proved. □

4.1.1.3 Brief Discussion on Stable Systems

A few words about the application of the latter method to stable systems can be here useful. If $a_1, a_2 < 0$, following the procedure used for the unstable case one finds

$$\frac{k_2}{k_1} = \frac{a_2\sqrt{a_2^2+4a_1}}{2a_1}. \tag{4.14}$$

In order to have wellposedness of the previous relation one should require $a_2^2 + 4a_1 > 0$; note that this condition forces the state matrix A to have real eigenvalues. Choose $k_1 = -\eta$, $k_2 = -\varepsilon$, that is

$$\eta = \frac{2\varepsilon a_1}{a_2\sqrt{a_2^2+4a_1}} > 0.$$

Set $\mathbf{K} = \mathbf{K}(\varepsilon) = -\varepsilon(\frac{2a_1}{a_2\sqrt{a_2^2+4a_1}}, 1)$. Application of the linear feedback $u = \langle \mathbf{K}(\varepsilon), \mathbf{x} \rangle$ to (4.1) with $a_1, a_2 < 0$ improves system stability. Moreover, if we take the initial condition $\mathbf{x}_0 \in \{x \in \mathbb{R}^2 : |\langle \mathbf{K}(\varepsilon), \mathbf{x} \rangle| < M\}$, the control variable satisfies $|u| < M$ for any $t > 0$. If $a_1 = 0$ and $a_2 < 0$ the same result holds with $\mathbf{K}$ given by

$$k_1 = a_2\varepsilon,\; k_2 = -\varepsilon < 0.$$

In particular $\mathbf{K}$ turns out to belong to $\ker(\mathbf{A})$. It should be emphasized that in the stable case one has $\|\mathbf{K}(\varepsilon)\| = O(\varepsilon)$ therefore, letting ε go to zero, one can make the width of the strip $\{\mathbf{x} \in \mathbb{R}^2 : |\langle \mathbf{K}(\varepsilon), \mathbf{x} \rangle| < M\}$ arbitrarily large. In other words semiglobal asymptotic stabilization by bounded control is achieved.

4.1.2 A Simulation Example

Consider the continuous-time system reported in [44], [45] setting the time-delay constant $\tau = 0$. Up to this assumption the system matrices $\mathbf{A}$ and $\mathbf{b}$ are given by

$$\mathbf{A} = \begin{pmatrix} 1 & 1.5 \\ 0.3 & -2 \end{pmatrix} + \begin{pmatrix} 0 & -1 \\ 0 & 0 \end{pmatrix} \quad \mathbf{B} = \begin{pmatrix} 10 \\ 1 \end{pmatrix} \tag{4.15}$$

The pair $(\mathbf{A}, \mathbf{b})$ is controllable and its canonical form is

$$\overline{\mathbf{A}} = \begin{pmatrix} 0 & 1 \\ 2.15 & -1 \end{pmatrix} \quad \overline{\mathbf{B}} = \begin{pmatrix} 0 \\ 1 \end{pmatrix} \tag{4.16}$$

The eigenvalues of the system are $\lambda_1 = -0.5 - 2\sqrt{0.6}$ and $\lambda_2 = -0.5 + 2\sqrt{0.6}$ with $\lambda_2 > 0$ and $\lambda_1 + \lambda_2 < 0$. According to Lemma 4.1, taking

$$\varepsilon = 2 > \frac{\sqrt{a_2^2 + 4a_1} - a_2}{2} \approx 1.05$$

we get

$$\mathbf{K} = (-1 - 4\sqrt{0.6}, -2).$$

Having fixed the saturation level to $M = 1$, we have simulate the closed-loop system evolution with initial state $\mathbf{x}_0 = (5, -2.99 - 2\sqrt{15})$ and control input $u(t) = \langle \mathbf{K}, \mathbf{x}(t) \rangle$. Note that $\|x_0\| \approx 11.8$ and $\langle \mathbf{K}, \mathbf{x}_0 \rangle \approx 0.98$, so that $u(0)$ is very close to the saturation level M. A plotting of the evolution of the system variables during the time interval $(0,5)$ is shown in Fig. 4.1-4.3. The region of attraction associated to $u(t)$ in the presence of saturation on the actuator is the unbounded strip

$$E = \left\{ (x_1, x_2) \in \mathbb{R}^2 : |x_1 + 4\sqrt{0.6}x_1 + 2x_2| < 1 \right\}.$$

The closed-loop state matrix, corresponding to the linear feedback without saturation, is given by

$$\overline{\mathbf{A}} + \overline{\mathbf{B}}\mathbf{K}^T = \begin{pmatrix} 0 & 1 \\ 1.15 - 4\sqrt{0.6} & -3 \end{pmatrix}.$$

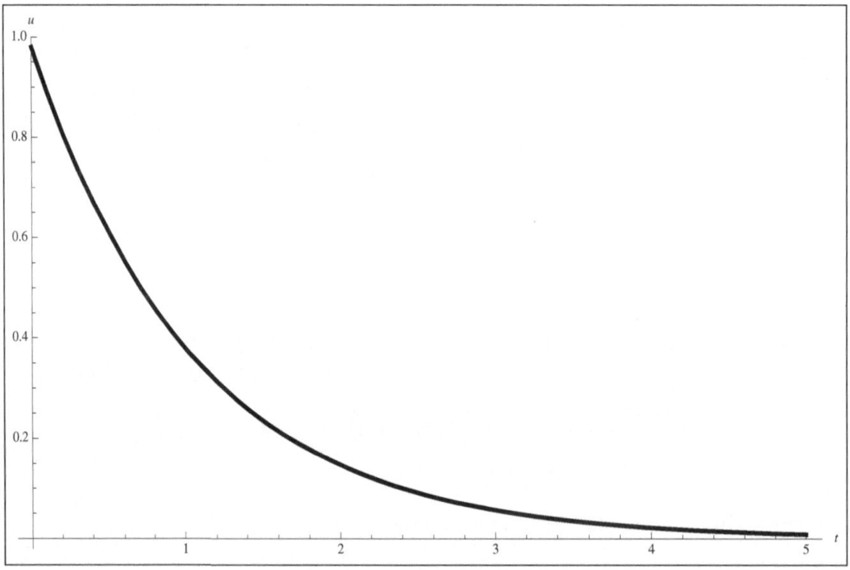

Fig. 4.1. Control variable $u(t)$.

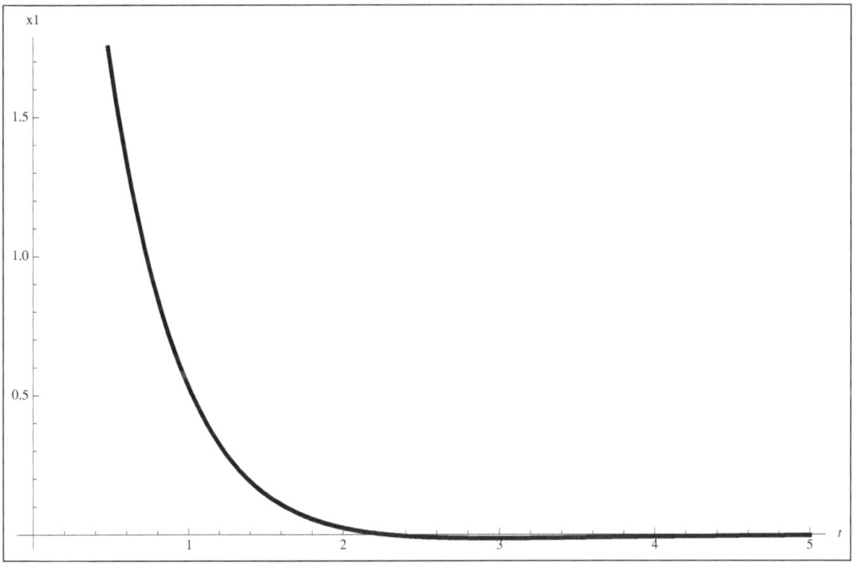

Fig. 4.2. State variable $x_1(t)$.

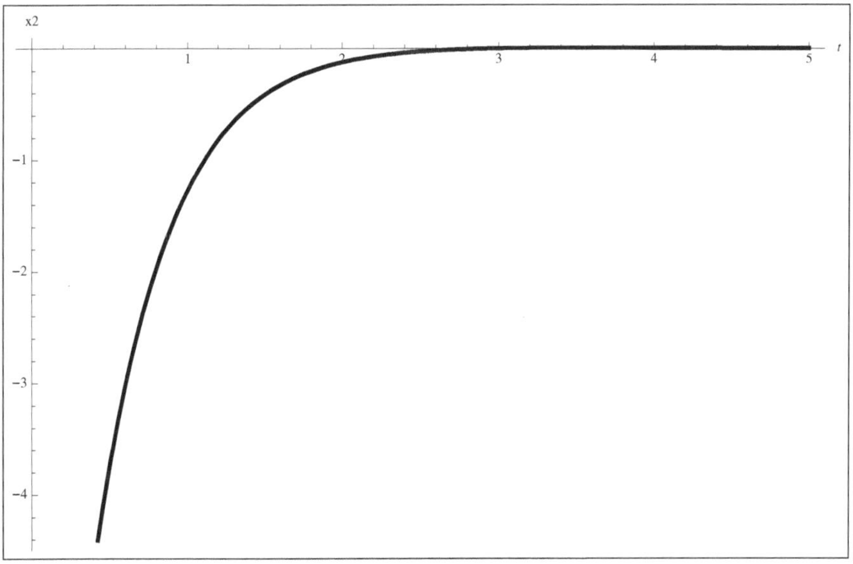

Fig. 4.3. State variable $x_2(t)$.

4.1.3 Multidimensional Systems

Consider a n-dimensional ($n \geq 2$) controllable SISO linear system having, without loss of generality, the following structure

$$\begin{cases} \dot{\mathbf{x}} = \mathbf{A}\mathbf{x} + \mathbf{B}u \\ \\ \mathbf{x}(0) = \mathbf{x}_0 \end{cases} \tag{4.17}$$

where

$$\mathbf{A} = \begin{pmatrix} 0 & 1 & \cdots & 0 \\ \vdots & \vdots & \vdots & \vdots \\ 0 & \cdots & 0 & 1 \\ a_1 & a_2 & \cdots & a_n \end{pmatrix}, \qquad \mathbf{B} = \begin{pmatrix} 0 \\ \vdots \\ 0 \\ 1 \end{pmatrix},$$

$\mathbf{x}(t) = (x_1(t), x_2(t), ..., x_n(t)) \in C^0([0,+\infty), \mathbb{R}^n)$ is the state vector and $v(t) \in C^0([0,+\infty), \mathbb{R})$ is the input variable.

Let us denote with $\{\lambda_1, \lambda_2, ..., \lambda_n\}$ the eigenvalues of A. We will suppose

$$\mathrm{Re}(\lambda_j) < 0 \text{ for all } j \leq n-1, \ \lambda_n \in [0,+\infty). \tag{4.18}$$

As a consequence, we have $a_1 \geq 0$ for any $n \in \mathbb{N}$. Input saturation is assumed in the system; in particular the input u is subject to the constraint

$$u = \mathbf{sat}_M(v), \ M > 0 \tag{4.19}$$

where $\mathbf{sat}_M(s)$ is given by (1.4) and $u(t) \in C^0([0,\infty), \mathbb{R})$.

Lemma 4.2. *Let be $\mathbf{K}^{(j)}$ an eigenvector associate to the real eigenvalue λ_j of the matrix $\mathbf{A}^T$, with $\mathbf{A}$ given by (4.17). For $\sigma \in \mathbb{R}$ we define the matrix $\mathbf{A}_\sigma^{(j)} = \mathbf{A} + \sigma \mathbf{b}(\mathbf{K}^{(j)})^T$; the eigenvalues of $\mathbf{A}_\sigma^{(j)}$ are*

$$\lambda_1, ..., \lambda_{j-1}, \lambda_j + \sigma k_n^{(j)}, ..., \lambda_n,$$

where $k_n^{(j)}$ is the nth component of $\mathbf{K}^{(j)}$.

Proof. The eigenvector $\mathbf{K}^{(j)}$ satisfies the following system of equations

$$a_1 k_n^{(j)} = \lambda_j k_1^{(j)}, \tag{4.20}$$

$$k_{s-1}^{(j)} + a_s k_n^{(j)} = \lambda_j k_s^{(j)} \quad \forall\, 1 < s \leq n; \tag{4.21}$$

we have denoted by $k_s^{(j)}$ the sth component of $\mathbf{K}^{(j)}$, $1 \leq s \leq n$. With a recursive procedure from equations (4.21) we can derive an equivalent system given by

$$k_s^{(j)} = k_n^{(j)} \left(\mu_j^{n-s} - \sum_{r=s}^{n-1} \mu_j^{r-s} a_{r+1} \right) \quad \forall\, 1 \leq s < n. \tag{4.22}$$

Consider the polynomial $q(\lambda) = \prod_{r \neq j}(\lambda - \lambda_r) = \lambda^{n-1} + \sum_{r=0}^{n-2} b_{r+1}\lambda^r$. Recalling the explicit expressions of the coefficients a_r and b_r in terms of the eigenvalues $\{\mu_r\}_{r=1}^n$, we see that

$$\mu_j^{n-s} - \sum_{r=s}^{n-1} \mu_j^{r-s} a_{r+1} = b_s \quad \forall\, 1 \leq s < n. \tag{4.23}$$

The characteristic polynomial of the matrix $\mathbf{A}_t^{(j)}$ is given by

$$p_\sigma^{(j)}(\lambda) = \lambda^n - \sum_{r=0}^{n-1}\left(a_{r+1} + \sigma k_{r+1}^{(j)}\right)\lambda^r =$$

$$= \lambda^n - \sum_{r=0}^{n-1} a_{r+1}\lambda^r - \sigma k_n^{(j)}\left(\lambda^{n-1} + \sum_{s=1}^{n-2}\left(\mu_j^{n-s} - \sum_{r=s}^{n-1}\mu_j^{r-s}a_{r+1}\right)\lambda^{s-1}\right),$$

where we have used the equations (4.22); by formula (4.23) we get

$$p_\sigma^{(j)}(\lambda) = \prod_{r=1}^{n}(\lambda - \lambda_r) - \sigma k_n^{(j)} q(\lambda) = \left((\lambda - \mu_j) - \sigma k_n^{(j)}\right) q(\lambda).$$

We obtain that equation $p_\sigma^{(j)}(\lambda) = 0$ is satisfied for $\lambda = \mu_r$, with $r \neq j$, and for $\lambda = \mu_j + \sigma k_n^{(j)}$. □

Lemma 4.3. *Given the system (4.17) with the linear feedback $u = \langle \mathbf{K}, \mathbf{x} \rangle$, the control norm $|u|$ is non-increasing along the closed-loop system trajectories if and only if* $\mathbf{K}$ *is an eigenvector of $\mathbf{A}^T$ with $\lambda < -k_n$, where $\lambda \in \mathbb{R}$ is the corresponding eigenvalue and k_n is the nth component of* $\mathbf{K}$.

Proof. Non-increasing property of $|u|$ can be written as

- if $\langle \mathbf{K}, \mathbf{x} \rangle \geq 0$ then $\langle \mathbf{K}, \dot{\mathbf{x}} \rangle \leq 0$,
- if $\langle \mathbf{K}, \mathbf{x} \rangle \leq 0$ then $\langle \mathbf{K}, \dot{\mathbf{x}} \rangle \geq 0$.

Given $M > 0$, we consider the level set $\langle \mathbf{K}, \mathbf{x} \rangle = M$; our aim is to construct $\mathbf{K}$ such that $\langle \mathbf{K}, \mathbf{A}\mathbf{x} + \mathbf{B}u \rangle \leq 0$ over the previous set, that is

$$\langle \mathbf{K}, \mathbf{A}\mathbf{x} + M\mathbf{B} \rangle \leq 0. \tag{4.24}$$

The two affine subspaces $\langle \mathbf{K}, \mathbf{x} \rangle = M$ and $\langle \mathbf{K}, \mathbf{A}\mathbf{x} + M\mathbf{B} \rangle = 0$ must be parallel and this can hold if and only if $\mathbf{K}$ is an eigenvector of the matrix $\mathbf{A}^T$. Denoting with $\lambda \in \mathbb{R}$ the eigenvalue associate to $\mathbf{K}$, for $\langle \mathbf{K}, \mathbf{x} \rangle = M$ we have

$$\langle \mathbf{K}, \mathbf{A}\mathbf{x} \rangle = M\lambda, \quad \langle \mathbf{K}, M\mathbf{B} \rangle = Mk_n$$

We see that condition (4.24) holds if the eigenvalue λ satisfies $\lambda \leq -k_n$. Parallelism between the spaces $\langle \mathbf{K}, \mathbf{x} \rangle = -M$ and $\langle \mathbf{K}, \mathbf{A}\mathbf{x} - M\mathbf{B} \rangle = 0$ gives the same condition on $\mathbf{K}$. This concludes the proof. □

Theorem 4.2. *Given the system (4.17) satisfying condition (4.18), there exists a vector* $\overline{\mathbf{K}} \in \mathbb{R}^n$ *such that for any* $M > 0$

1. $\Omega_M = \{\mathbf{x} \in \mathbb{R}^n : |\langle \overline{\mathbf{K}}, \mathbf{x} \rangle| \leq M\}$ *is an invariant set for the solution* $\phi(t, \mathbf{x}_0, \overline{\mathbf{K}})$*;*
2. *if* $\mathbf{x}_0 \in \Omega_M$ *we have* $\lim_{t\to\infty} \|\phi(t, \mathbf{x}_0, \overline{\mathbf{K}})\| = 0$ *under the saturation constraint (4.19).*

Proof. Let $\mathbf{K}^{(n)}$ be an eigenvector of $\mathbf{A}^T$, associated to the positive eigenvalue μ_n and suppose without loss of generality $k_n^{(n)} > 0$. We set $\overline{\mathbf{K}} = \sigma \mathbf{K}^{(n)}$, with $\sigma \in \mathbb{R}$ to be determined. Accordingly to Lemma 4.3, the set $\Omega_M = \{\mathbf{x} \in \mathbb{R}^n : |\langle \overline{\mathbf{K}}, \mathbf{x} \rangle| \leq M\}$ is invariant for the solution if we take

$$\sigma \leq -\frac{\mu_n}{k_n^{(n)}}.$$

Moreover, by Lemma 4.2, the eigenvalues of the closed-loop system $\{\overline{\mu}_j\}$ are given by

$$\overline{\mu}_j = \mu_j \quad \forall\, 1 \leq j < n,$$
$$\overline{\mu}_n = \mu_n + \sigma k_n^{(n)}.$$

Recall that, by assumption, $Re(\mu_j) < 0$ for any $1 \leq j < n$. Choosing

$$\sigma < -\frac{\mu_n}{k_n^{(n)}},$$

we obtain asymptotic stability under the feedback $u = \langle \overline{\mathbf{K}}, \mathbf{x} \rangle = \langle \sigma \mathbf{K}^{(n)}, \mathbf{x} \rangle$. The Theorem is proved. □

4.1.4 Semiglobal Stabilization by a Finite Number of Actuators

The aim of this section is to introduce a battery of singularly saturating actuators, with predeterminable cardinality, such that taking their sum we reconstruct the linear feedback $\langle \mathbf{K}, \mathbf{x} \rangle$ ensuring semiglobal asymptotic stability of the closed-loop system. To this purpose we modify the saturation constraint as follows

$$u = \sum_j u_j \text{ where } u_j = \text{sat}_M(v_j), \text{ with } M > 0 \text{ and} \tag{4.25}$$

$$v_j(t) \in C^0([0, +\infty), \mathbb{R}), \; j \in \mathbb{N}.$$

Denoting by $\mathscr{D}_M$ the domain of attraction associated to $\langle \mathbf{K}, \mathbf{x} \rangle$ with saturation level fixed to $M > 0$, if we assume $j = 1, ..., r$ in (4.25) then the domain of attraction associated to u is $\mathscr{D}_{rM}$.

In Section 4.1.3 we have shown how to design a stabilizing linear feedback decreasing (in norm) along the closed-loop system trajectory. Thanks to the invariance of the level sets of such feedback, we can state the following result.

Theorem 4.3. *Assume the saturation constraint (4.25) on system (4.17) and suppose condition (4.18) holding; consider the linear feedback* $\langle \overline{\mathbf{K}}, \mathbf{x} \rangle$*, where the vector* $\overline{\mathbf{K}}$ *is given by Theorem 4.1. For any fixed* $M > 0$*,* $\delta > 0$ *arbitrarely small and* $\mathbf{x}_0 \in \mathbb{R}^n$ *there exists* $r = r(\mathbf{x}_0, \overline{\mathbf{K}}, M, \delta) \in \mathbb{N}$ *such that we can find a set of r functions* $\{f_j(\mathbf{x})\}_{j=1}^r$ *with the following properties:*

1. $\sup_{\mathbf{x}\in\mathbb{R}^n} |f_j(\mathbf{x})| < M - \delta \; \forall j = 1,..,r$*;*
2. *Setting* $u(\mathbf{x}) = \sum_{j=1}^r f_j(\mathbf{x}) = \sum_{j=1}^r \mathrm{sat}_M(f_j(\mathbf{x}))$*, we have* $u(\phi(t,\mathbf{x}_0)) = \langle \overline{\mathbf{K}}, \phi(t,\mathbf{x}_0) \rangle$*.*

In particular $r = \left[\dfrac{|\langle \overline{\mathbf{K}}, \mathbf{x}_0 \rangle|}{M-\delta} \right] + 1$*, where* $[s]$ *stands for the integer part of the number s.*

Proof. In combination with the truncation operator (1.1) we consider the remainder

$$G_h(s) = s - \mathrm{sat}_h(s)$$

Given the system (4.17) under assumption (4.18) we take $\mathbf{K}$ as designed in Lemma 4.2 and Lemma 4.3. Let us fix a real number $\delta > 0$ arbitrarily small and define the following partition of $\mathbb{R}^n$:

$$\mathbb{R}^n = \bigcup_{j=1}^{\infty} \Omega_j^{\delta}$$

with

$$\Omega_j^{\delta} = \{\mathbf{x} \in \mathbb{R}^n : (j-1)(M-\delta) \le |\langle \mathbf{K}, \mathbf{x} \rangle| < j(M-\delta)\}, \tag{4.26}$$

where M is the saturation level of the actuators. Note that the level sets of the linear function $\langle \mathbf{K}, \mathbf{x} \rangle$ are hyperplanes; this implies that any Ω_j^{ε} consists of a pair of unbounded strips.

Let us define the sequence of functions $\{f_j\}$, $j \ge 1$

$$f_j(\mathbf{x}) = \mathrm{sat}_{M-\delta}(G_{j(M-\delta)}(\langle \mathbf{K}, \mathbf{x} \rangle)).$$

It is easy to verify that $\|f_j(\mathbf{x})\| \le M - \delta$ for any j.

Fix the initial condition $\mathbf{x}_0 \in \mathbb{R}^n$; we have $\mathbf{x}_0 \in \Omega_r^{\delta}$ for some $r \in \mathbb{N}$. In particular, recalling that for any $\mathbf{x} \in \Omega_j^{\delta}$ we have

$$j - 1 \le \frac{\langle \mathbf{K}, \mathbf{x} \rangle}{M-\delta} < j,$$

r is given by

$$r = \left[\frac{|\langle \mathbf{K}, \mathbf{x} \rangle|}{M-\delta} \right] + 1.$$

Setting

$$u(\mathbf{x}) = \mathrm{sat}_{M-\delta}(\langle \mathbf{K}, \mathbf{x} \rangle) + \sum_{j=1}^{r} f_j(\mathbf{x}), \tag{4.27}$$

by construction it verifies

$$u(\mathbf{x}) = \langle \mathbf{K}, \mathbf{x} \rangle \text{ for any } \mathbf{x} \in \bigcup_{j=1}^{r} \Omega_j^{\delta}.$$

Recalling that under the feedback $\langle \mathbf{K}, \mathbf{x} \rangle$ the sets Ω_C, $C > 0$, are invariant for the solution of the closed-loop system in the presence of saturation constraint (2.2) and that $\mathbf{K}$ has been designed in order to guarantee the asymptotic stability of the system, Theorem 4.3 is proved. □

Remark 4.2. *Note that if the solution $\phi(t, \mathbf{x}_0, u)$ lies in $\Omega_{j_0}^{\delta}$ for some j_0 then $f_j(\phi(t, \mathbf{x}_0, u)) \equiv 0$ for any $j > j_0$. In particular any $f_j(\phi(t, \mathbf{x}_0, u))$ with $j > 1$ is different from zero at most as t varies in a bounded interval.*

Remark 4.3. *Since the level set $\langle \mathbf{K}, \mathbf{x} \rangle = M$ lies in the semispace $\langle \mathbf{K}, \mathbf{A}\mathbf{x} + M\mathbf{b} \rangle < 0$, if the initial condition $\mathbf{x}_0$ belongs to the strip*

$$\left\{ r(M-\delta) \leq \langle \mathbf{K}, \mathbf{x} \rangle < r(M-\delta)\frac{|k_1|}{a_1} \right\} \subset \Omega_{r+1}^{\delta},$$

we have asymptotic stabilization under the feedback

$$u = \mathrm{sat}_{M-\delta}(\langle \mathbf{K}, \mathbf{x} \rangle) + \sum_{j=1}^{r} f_j(\mathbf{x}).$$

Note that $(a_1 k_2 / k_1) < -k_2$ is the positive eigenvalue of the state matrix $\mathbf{A}$.

Remark 4.4. *The real parameter $\delta > 0$ has been introduced to saturate each actuator on a level which is less than the maximum one. This has been done in order to prevent an eventual fault of the plant due to overload of the input devices.*

4.1.5 A Planar Example

Let us set the saturation level to $M = 2$ and define the state matrix $\mathbf{A}$ choosing the parameters $a_1 = 3, a_2 = 2$, that is

$$\mathbf{A} = \begin{pmatrix} 0 & 1 \\ 3 & 2 \end{pmatrix}.$$

With this choice the eigenvalues of $\mathbf{A}$ are $\mu_1 = -1, \mu_2 = 3$. An eigenvector of $\mathbf{A}^T$ associated to μ_2 is given by $\mathbf{K} = (\sigma, \sigma)$, with $\sigma \in \mathbb{R} \setminus \{0\}$. In order to have asymptotic stability of the system driven by the feedback $u = \langle \mathbf{K}, \mathbf{x} \rangle$, we need to impose $\sigma < -\mu_2 = -3$; for sake of simplicity we take $\sigma = -4$. The eigenvalues of the closed-loop matrix $\overline{\mathbf{A}} = \mathbf{A} + \mathbf{b}\mathbf{K}^T$ are $\overline{\mu}_1 = \overline{\mu}_2 = -1$.

The initial condition $\mathbf{x}_0 = (3/2, 3/4)$ is considered, and the saturation constraint (4.25) on the control variable is assumed; it is easy to verify that $\langle \mathbf{K}, \mathbf{x}_0 \rangle = -9$ therefore, according to (4.26), one has $\mathbf{x}_0 \in \Omega_5^\delta$ for $\delta \in \mathbb{R}$ sufficiently small. Figure 4.4 shows the evolution of the system in the (x_1, x_2)-plane for the time interval $[0,7]$; the straight lines are the boundary components of the sets Ω_j^δ with $\delta = 0$. Figure 4.5 represents the control $u(t)$ as t varies in the interval $[0,7]$. Since the norm of the control $v = \sum_{j=1}^{5} \mathrm{sat}_{M-\delta}(G_{j(M-\delta)}(\langle \mathbf{K}, \mathbf{x} \rangle))$ is decreasing by construction, the number of employed actuators reduces from five to one during the system evolution; in particular any intersection of the state $\mathbf{x}(t)$ with the boundary lines of the strips corresponds to the deactivation of one actuator (see Figure 4.4). Note that for $t = 7$ both $\|\mathbf{x}(t)\|$ and $|v(t)|$ are close to the origin: we have

$$\|\mathbf{x}(7)\| = \frac{3\sqrt{929}}{4e^7} \approx 0.0208453 \quad \text{and} \quad |v(7)| = \frac{9}{e^7} \approx 0.00820694.$$

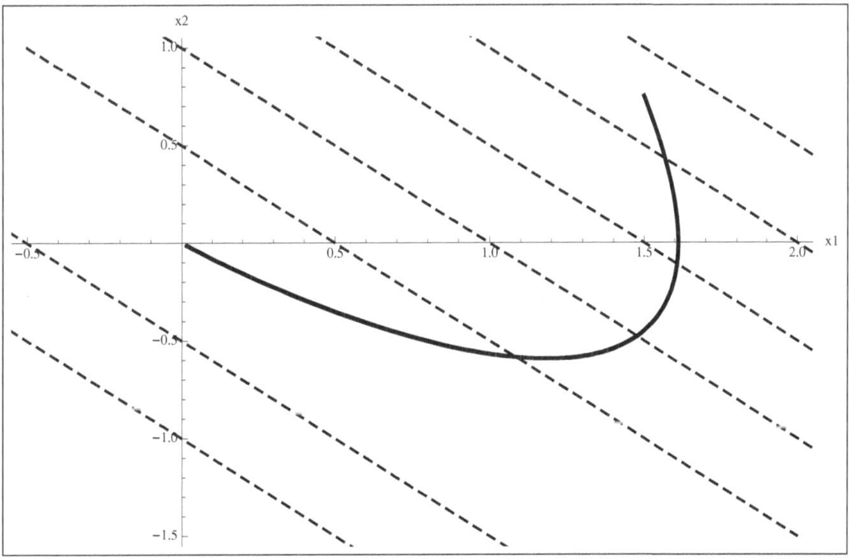

Fig. 4.4. Evolution of $(x_1(t), x_2(t))$ for $t \in [0,7]$.

4.1.6 Extension to Multi-Input Systems

Let us consider a n-dimensional multi-input (MI) linear system having the following structure

$$\begin{cases} \dot{\mathbf{x}} = \mathbf{A}\mathbf{x} + \mathbf{B}\mathbf{u} \\ \mathbf{x}(0) = \mathbf{x}_0 \end{cases} \tag{4.28}$$

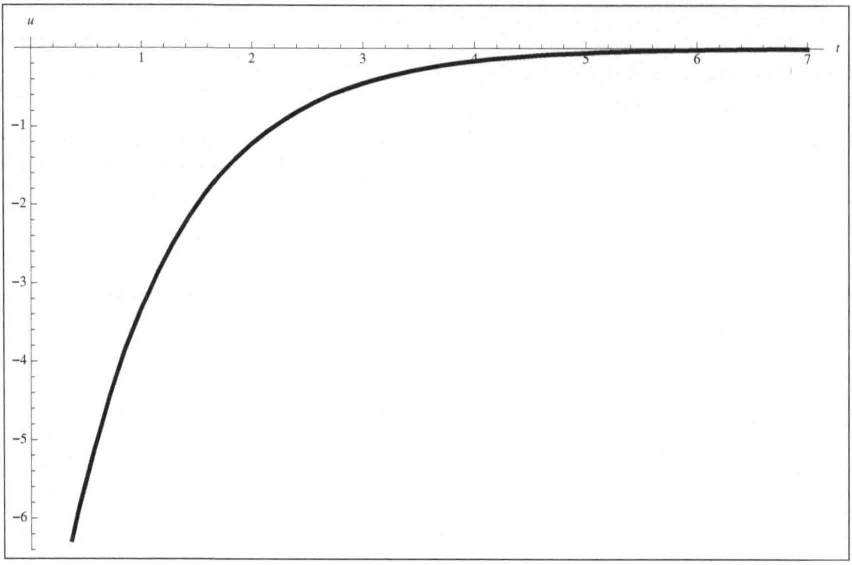

Fig. 4.5. Control variable $u(t)$ for $t \in [0,7]$.

where $\mathbf{A} \in \mathscr{M}^{n\times n}$, $\mathbf{B} \in \mathscr{M}^{n\times m}$, $\mathbf{x}(t) \in C^0(\mathbb{R},\mathbb{R}^n)$ is the state vector and $\mathbf{u}(t) \in L^\infty(\mathbb{R},\mathbb{R}^m)$ is the input variable. For an arbitrary square matrix $\tilde{\mathbf{A}} \in \mathscr{M}^{n\times n}$, the spectrum of $\tilde{\mathbf{A}}$, i.e. the set of eigenvalues, is indicated with $\sigma(\tilde{\mathbf{A}}) \subset \mathbb{C}$; in addition, set $\mathbb{R}^{m,+} = \{\mathbf{x} \in \mathbb{R}^m : x_i > 0, i = 1,...,m\}$ and $\mathbb{C}^- = \{\omega \in \mathbb{C} : \mathrm{Re}(\omega) < 0\}$.

In this section it is discussed the problem of extending to multi input systems the results on the existence of controlled invariant strips. To this purpose the following general definition can be useful.

Definition 4. The control feedback $\mathbf{u} = \mathbf{u}(\mathbf{x}(t)) = (u_1(\mathbf{x}(t)),...,u_m(\mathbf{x}(t)))$ is said to satisfy the *Sublevel Sets Invariance and Stabilization* property (SSIS) if the following conditions hold:

1. System (4.28) driven by $\mathbf{u}(\mathbf{x}(t))$ is asymptotically stable.
2. For any $\mathbf{M} = (M_1,...,M_m) \in \mathbb{R}^{m,+}$, the set

$$E_{\mathbf{M}} = \{\mathbf{x} \in \mathbb{R}^n : |u_i(\mathbf{x})| < M_i,\ i = 1,...,m\} \tag{4.29}$$

is a controlled invariant set for the solution of System (3.11) driven by $\mathbf{u}(\mathbf{x}(t))$.

Remark 4.5. *The SSIS condition means asymptotic stability of the closed-loop system together with the invariance of the control norm sublevel sets; it is a generalization of the invariance condition presented in Section 4.1.3 for single input systems.*

There is a strict connection between the structure of the saturated maximal region of attraction for a system (or null controllable region) and the existence of SSIS controls. It follows from the definition that if $\mathbf{u}(t)$ is a SSIS control, then the sublevel set $E_{\mathbf{M}}$ (defined in (4.29)) verifies

$$E_{\mathbf{M}} \subseteq \mathscr{B}_{\mathbf{M}} \quad \forall\, \mathbf{M} \in \mathbb{R}^{m,+}. \tag{4.30}$$

The existence of SISS controls will be proved for special classes of systems first and then some general conditions of existence will be presented.

4.1.6.1 Systems with a Single Unstable Mode

This section deals with the problem of stabilizing a controllable MI system having a single unstable real mode. In particular, without loss of generality, it is assumed that

$$\mathrm{Re}(\nu_j) < 0 \quad \forall\, 1 \le j \le n-1,$$

$$\nu := \nu_n \ge 0,$$

where $\{\nu_j\}_{j=1}^{n} = \sigma(\mathbf{A})$. The main aim is to design a linear feedback able to stabilize semiglobally the system on the saturated maximal region of attraction. Note that, since there is only one unstable direction, in this case the saturated maximal region of attraction is a region bounded by two parallel $(n-1)$ dimensional hyperplanes.

Let $\mathbf{P} \in \mathscr{M}^{n \times n}$ the matrix associated to the following change of coordinates (Jordan canonical decomposition):

$$\overline{\mathbf{A}} = \mathbf{P}^{-1}\mathbf{A}\mathbf{P}, \quad \overline{\mathbf{B}} = \mathbf{P}^{-1}\mathbf{B}$$

with

$$\overline{\mathbf{A}} = \begin{pmatrix} \mathbf{A}_{11} & \mathbf{0} \\ & \\ \mathbf{0} & \nu \end{pmatrix} \qquad \overline{\mathbf{B}} = \begin{pmatrix} \overline{\mathbf{B}}_1 & \cdots & \overline{\mathbf{B}}_m \\ & & \\ b_1 & \cdots & b_m \end{pmatrix}$$

The matrix $\mathbf{A}_{11} \in \mathscr{M}^{(n-1)\times(n-1)}$ has stable eigenvalues by definition. The new state variables are $\mathbf{z} = (\mathbf{z}_1, z_n) := (z_1, \ldots, z_{n-1}, z_n)$ with $\mathbf{z} = \mathbf{P}^{-1}\mathbf{x}$. Due to plant controllability assumption, at least one of the coefficients b_j, $j = 1, \ldots, m$, is different from zero. Without loss of generality assume that

$$b_j \ne 0 \,\forall\, 1 \le j \le s \quad \text{with } s \le m.$$

Lemma 4.4. *For any $\varepsilon > 0$ and for $1 \le j \le s$, the linear feedback*

$$\mathbf{u}_{\varepsilon}^{(j)}(\mathbf{z}) = -\frac{1}{b_j}(\overbrace{0, \ldots, 0}^{j-1}, (\nu + \varepsilon) z_n, \overbrace{0, \ldots, 0}^{m-j})$$

ensures asymptotic stabilization of the saturating plant $(\overline{\mathbf{A}}, \overline{\mathbf{B}}, \mathbf{M})$ *on the region*

$$E_\varepsilon^{(j)} = \left\{ (\mathbf{z}_1, z_n) \in \mathbb{R}^n : |z_n| \leq \frac{M_j |b_j|}{\nu + \varepsilon} \right\}.$$

Proof. It is straightforward to verify that the control $\mathbf{u}_\varepsilon^{(j)}(\mathbf{z})$ satisfies the saturation constraints for any state $\mathbf{z} \in E_\varepsilon^{(j)}$. Let us denote by $\mathbf{F}^{(j)} \in \mathcal{M}^{m \times n}$ the matrix whose entries $\left\{ f_{ik}^{(j)} \right\}$ satisfy

$$f_{ik}^{(j)} = \delta_{ij} \delta_{kn},$$

where δ_{ij} are the standard Kronecker symbols. The asymptotic stability of the plant follows observing that the closed-loop system is given by

$$\dot{\mathbf{z}} = (\overline{\mathbf{A}} - (\nu + \varepsilon) \overline{\mathbf{B}} \mathbf{F}^{(j)}) \mathbf{z},$$

whose eigenvalues are

$$\sigma(\overline{\mathbf{A}} - (\nu + \varepsilon) \overline{\mathbf{B}} \mathbf{F}^{(j)}) = \sigma(\mathbf{A}_{11}) \cup \{-\varepsilon\} \subset \mathbb{C}^-.$$

□

The lemma shows that the system can be stabilized using separately s different SSIS controllers, where s is the number of non-null coefficients b_j; next result illustrates how to obtain semiglobal asymptotic stabilization over the saturated maximal region of attraction. Let us define the set

$$E_{max} = \left\{ \mathbf{z} \in \mathbb{R}^n : |z_n| < \frac{1}{\nu} \sum_{i=1}^{s} M_i |b_i| \right\}.$$

Theorem 4.4. *The set* E_{max} *is the saturated maximal region of attraction for the system* $(\overline{\mathbf{A}}, \overline{\mathbf{B}}, \mathbf{M})$*; moreover for any* $\mathbf{z}_0 \in E_{max}$ *there exists a linear feedback* $\mathbf{u}_{\mathbf{z}_0}$ *ensuring asymptotic stabilization of the plant.*

Proof. Suppose that $\mathbf{z}_0 = (z_{1,0}, \ldots, z_{n,0}) \notin E_{max}$; without loss of generality assume that

$$z_{n,0} \geq \frac{1}{\nu} \sum_{i=1}^{s} M_i |b_i|. \tag{4.31}$$

Now, for an arbitrary control input u

$$\dot{z}_n = \nu z_n + \sum_{i=1}^{m} b_i u_i(t).$$

For plant stabilization, since $z_n(0) > 0$ by (4.31), it is required

$$\dot{z}_n(0) < 0,$$

that is equivalent to

$$\nu z_{n,0} < -\sum_{i=1}^{m} b_i u_i(t) \leq \sum_{i=1}^{s} M_i |b_i|.$$

This condition is incompatible with assumption (4.31). It has been proved proved that, if the initial data are choosen outside the set E_{max}, there exists no admissible control input that can ensure plant stability.

Let us define the control vector $\mathbf{u}_\varepsilon = \mathbf{u}_{\varepsilon(\mathbf{z}_0)} = (u_{\varepsilon,1}, ..., u_{\varepsilon,m})$ as follows.

$$u_{\varepsilon,1}(\mathbf{z}) = \frac{-(\nu+\varepsilon)z_n}{b_1}$$

$$u_{\varepsilon,2}(\mathbf{z}) = \frac{b_1}{b_2}\left(\frac{-(\nu+\varepsilon)z_n}{b_1} - T_{M_1}\left(u_{\varepsilon,1}(\mathbf{z})\right)\right)$$

$$\vdots$$

$$u_{\varepsilon,s}(\mathbf{z}) = \frac{b_{s-1}}{b_s}\left(\frac{-(\nu+\varepsilon)z_n}{b_{s-1}} - T_{M_{s-1}}(u_{\varepsilon,s-1}(\mathbf{z}))\right)$$

$$u_{\varepsilon,j}(\mathbf{z}) \equiv 0 \;\forall\; s < j \leq m.$$

It is easy to verify that the eigenvalues of the closed-loop system are still $\sigma(\mathbf{A}_{11}) \cup \{-\varepsilon\}$. Moreover, the saturation constraints are fulfilled over the set

$$E_\varepsilon = \left\{ \mathbf{z} \in \mathbb{R}^n : |z_n| < \frac{1}{\nu+\varepsilon}\sum_{i=1}^{s} M_i |b_i| \right\}.$$

The conclusion follows observing that

$$\lim_{\varepsilon \to 0^+} E_\varepsilon = E_{max}.$$

□

4.1.6.2 Eigenstructured Systems

Let us discuss now the case of exactly $m \leq n$ real unstable modes. Without loss of generality it can be assumed that $\mathrm{Re}(\nu_j) < 0$ for any $j = 1, ..., n-m$ and $\nu_j \geq 0$ for $j = n-m+1, ..., n$. For sake of semplicity, set

$$\lambda_j = \nu_{n-m+j}, \quad j = 1, ..., m.$$

The system can be easily splitted into a stable/unstable subsystems partition as follows:

$$\overline{\mathbf{A}} = \begin{pmatrix} \mathbf{A}_s & \mathbf{0} \\ \mathbf{0} & \mathbf{A}_u \end{pmatrix} \qquad \overline{\mathbf{B}} = \begin{pmatrix} \mathbf{B}_s \\ \mathbf{B}_u \end{pmatrix},$$

where $\mathbf{A}_s \in \mathscr{M}^{(n-m)\times(n-m)}$, $\mathbf{A}_u \in \mathscr{M}^{m\times m}$, $\mathbf{B}_s \in \mathscr{M}^{(n-m)\times m}$, $\mathbf{B}_u \in \mathscr{M}^{m\times m}$ and

$$\overline{\mathbf{A}} = \mathbf{R}^{-1}\mathbf{A}\mathbf{R}, \quad \overline{\mathbf{B}} = \mathbf{R}^{-1}\mathbf{B}$$

for a suitable invertible matrix $\mathbf{R} \in \mathscr{M}^{n\times n}$. Let us focus the analysis on the unstable subsystem $(\mathbf{A}_u, \mathbf{B}_u)$.

Proposition 4.1. *If* $\mathbf{A}_u\mathbf{B}_u = \mathbf{B}_u\mathbf{A}_u$ *and* $\mathrm{rank}(\mathbf{B}_u) = m$ *then the system* $(\mathbf{A}, \mathbf{B})$ *admits a family of SSIS linear feedbacks.*

Proof. Since $\mathbf{A}_u$ and $\mathbf{B}_u$ commute, they are simultaneously diagonalizable, i.e. there exists $\mathbf{S} \in \mathscr{M}^{m\times m}$ such that

$$\mathbf{S}^{-1}\mathbf{A}_u\mathbf{S} = \mathrm{diag}(\lambda_1, ..., \lambda_m) =: \Delta_{\mathbf{A}}$$

$$\mathbf{S}^{-1}\mathbf{B}_u\mathbf{S} = \mathrm{diag}(b_1, ..., b_m) =: \Delta_{\mathbf{B}}.$$

Denote by $\mathbf{z} = (\overline{\mathbf{z}}, z_1, ..., z_m)$ the new coordinates.

From the condition $\mathrm{rank}(\mathbf{B}_u) = m$ follows

$$b_1 \cdot b_2 \cdots b_m \neq 0.$$

It is straighforward to verify that for any $\varepsilon > 0$ the control

$$\mathbf{u}(\mathbf{z}(t)) = (u_1(\mathbf{z}(t)), ..., u_m(\mathbf{z}(t))) \qquad u_j(\mathbf{z}) = -\frac{(\lambda_j + \varepsilon)}{b_j}\mathbf{z}_j$$

satisfies the SSIS property with respect to system $(\Delta_{\mathbf{A}}, \Delta_{\mathbf{B}})$. Set

$$\Delta_\varepsilon = -\mathrm{diag}\left(\frac{\lambda_1 + \varepsilon}{b_1}, ..., \frac{\lambda_m + \varepsilon}{b_m}\right)$$

Consider now the system $(\mathbf{S}^{-1}\mathbf{A}_u\mathbf{S}, \mathbf{S}^{-1}\mathbf{B}_u)$. Define the control

$$\mathbf{u}(\mathbf{x}(t)) = \mathbf{S}\Delta_\varepsilon\mathbf{x}(t).$$

The validity of the SSIS condition follows observing that the system $(\mathbf{S}^{-1}\mathbf{A}_u\mathbf{S}, \mathbf{S}^{-1}\mathbf{B}_u)$ driven by the control $\mathbf{u}(\mathbf{x}(t))$ is equivalent to system $(\Delta_{\mathbf{A}}, \Delta_{\mathbf{B}})$ driven by the control $\Delta_\varepsilon\mathbf{x}(t)$. Finally, to obtain a SSIS control for the system $(\mathbf{A}, \mathbf{B})$ one can proceed as follows. Define the matrices $\mathbf{H} \in \mathscr{M}^{n\times n}$ and $\mathbf{K}_\varepsilon \in \mathscr{M}^{m\times n}$ as

$$\mathbf{H} = \begin{pmatrix} \mathbf{I}_{(n-m)\times(n-m)} & \mathbf{0} \\ \mathbf{0} & \mathbf{S} \end{pmatrix}$$

$$\mathbf{K}_\varepsilon = (\mathbf{0} \quad \mathbf{S}\Delta_\varepsilon).$$

The linear feedback $\mathbf{u}(\mathbf{x}(t)) = \mathbf{K}_\varepsilon \mathbf{x}(t)$ is a SSIS control for the system

$$(\mathbf{R}^{-1}\mathbf{H}^{-1}\mathbf{A}\mathbf{S}\mathbf{R}, \mathbf{R}^{-1}\mathbf{H}^{-1}\mathbf{B})$$

and the statement is proved. □

From the Proposition 4.1, it follows that the existence of SSIS controls for a general system $(\mathbf{A},\mathbf{B})$ with m unstable modes is ensured if one can find a linear transformation $\mathbf{R} \in \mathscr{M}^{n\times n}$ with both $\mathbf{R}^{-1}\mathbf{A}_u\mathbf{R}$ and $\mathbf{R}^{-1}\mathbf{B}_u\mathbf{R}$ diagonal.

Let us denote by $w_{\lambda_1},...,w_{\lambda_m}$ the eigenvectors of the matrix $\mathbf{A}$ associated to the non-negative eigenvalues $\lambda_1,...,\lambda_m$. Set $\Lambda^\perp = \{\mathrm{span}(w_{\lambda_1},...,w_{\lambda_m})\}^\perp$. For $j = 1,...,m$ let us define the subspaces

$$\Lambda_j = \Lambda^\perp + \mathrm{span}(w_{\lambda_j});$$

$$\Xi_j = \{\mathrm{span}(w_{\lambda_j})\}^\perp.$$

Proposition 4.2. *Let us denote by $\mathbf{B}_j$, $j = 1,...,m$ the column vectors of the matrix $\mathbf{B}$. If, up to an order rearrangement, for any $j = 1,...,m$ the following conditions hold*

$$\begin{cases} \mathbf{B}_j \in \Lambda_j \\ \mathbf{B}_j \notin \Xi_j, \end{cases} \tag{4.32}$$

then there exists a family of SSIS linear feedbacks for system $(\mathbf{A},\mathbf{B})$.

Proof. The vectors $\{w_{\lambda_j}\}$, $j = 1,...,m$ are independent and constitute a basis for the subspace Λ; take in addiction $n-m$ independent vectors $y_i \in \Lambda^\perp$. Since

$$\mathbb{R}^n = \Lambda \oplus \Lambda^\perp,$$

we have

$$\mathbb{R}^n = \mathrm{span}(y_1,...,y_{n-m},w_1,...,w_m).$$

Define the matrix $\mathbf{H} \in \mathscr{M}^{n\times n}$ as

$$\mathbf{H} = (y_1\ y_2 \cdots w_1 \cdots\ w_m).$$

Due to condition (4.32), the transformed matrices $\overline{\mathbf{A}} = \mathbf{H}\mathbf{A}\mathbf{H}^{-1}$, $\overline{\mathbf{B}} = \mathbf{H}\mathbf{B}$ have the following special form

$$\overline{\mathbf{A}} = \begin{pmatrix} \mathbf{A}_s & \mathbf{0} \\ \mathbf{0} & \Delta_\mathbf{A} \end{pmatrix} \qquad \overline{\mathbf{B}} = \begin{pmatrix} \mathbf{B}_s \\ \Delta_\mathbf{B} \end{pmatrix},$$

with $\Delta_\mathbf{A} = \mathrm{diag}(\lambda_1,...,\lambda_m)$ and $\Delta_\mathbf{B} = \mathrm{diag}(b_1,....,b_m)$, where

$$b_j = \langle \mathbf{B}_j, w_{\lambda_j}\rangle \neq 0\ \forall\ j = 1,...,m.$$

Since the unstable subsystem $(\Delta_{\mathbf{A}}, \Delta_{\mathbf{B}})$ turns out to be diagonal, the SSIS control can be derived as in the proof of Proposition 4.1. □

Remark 4.6. *Condition (4.32) can be achieved only if for any $j = 1,...,m$, there exists $s_j \leq (n-m+2)$ such that*

$$rank[\mathbf{B}_j\ \mathbf{A}\mathbf{B}_j \cdots \mathbf{A}^{s_j-1}\mathbf{B}_j] < s_j.$$

4.1.6.3 Examples

Example 4.1. Consider the following system $(\mathbf{A},\mathbf{B})$ with $n=5,\ m=4$:

$$\mathbf{A} = \begin{pmatrix} \frac{29}{5} & \frac{32}{5} & \frac{2}{5} & 2 & -\frac{12}{5} \\ -\frac{22}{5} & -\frac{21}{5} & -\frac{1}{5} & -1 & -\frac{4}{5} \\ -\frac{3}{5}+\frac{\sqrt{3}}{2} & \frac{7-5\sqrt{3}}{10} & \frac{1}{5}-\sqrt{3} & -2+\sqrt{3} & \frac{3}{10}+\sqrt{3} \\ -\frac{9}{5}-\frac{\sqrt{3}}{2} & \frac{-19+5\sqrt{3}}{10} & \frac{8}{5}+\sqrt{3} & -1-\sqrt{3} & \frac{9}{10}-\sqrt{3} \\ -\frac{12}{5} & -\frac{6}{5} & -\frac{1}{5} & -1 & -\frac{1}{5} \end{pmatrix}$$

and

$$\mathbf{B} = \begin{pmatrix} 1 & 0 & -\frac{1}{6} & 3 \\ 0 & 1 & 1 & -3 \\ -1 & 0 & 1 & 1 \\ 0 & -2 & -\frac{1}{2} & 0 \\ 3 & 0 & 1 & \frac{1}{3} \end{pmatrix}$$

The eigenvalues of the state matrix are $\nu_{1,2} = -12\mathrm{i}$, $\nu_{3,4} = -\sqrt{3}$ and $\nu_5 = 3$. By a coordinates change, the system can be rewritten as

$$\overline{\mathbf{A}} = \begin{pmatrix} -1 & -2 & 0 & 0 & 0 \\ 2 & -1 & 0 & 0 & 0 \\ 0 & 0 & 0 & 1 & 0 \\ 0 & 0 & -3 & -2\sqrt{3} & 0 \\ 0 & 0 & 0 & 0 & 3 \end{pmatrix}$$

and

$$\overline{\mathbf{B}} = \begin{pmatrix} 4 & 1 & \frac{11}{6} & \frac{1}{3} \\ 3 & -1 & 0 & \frac{10}{3} \\ 0 & 1 & -\frac{4}{3} & -1 \\ 9 & -5 & -\frac{13}{6} & \frac{14}{3} \\ -7 & 1 & 0 & 0 \end{pmatrix}.$$

Set

$$u_1(\mathbf{x}) = -\frac{3+\varepsilon}{-7}\mathbf{x}_5,$$

$$u_2(\mathbf{x}) = -7\left(-\frac{3+\varepsilon}{-7}\mathbf{x}_5 - T_{M_1}\left(-\frac{3+\varepsilon}{-7}\mathbf{x}_5\right)\right)$$

and

$$u_3 = u_4 = u_5 \equiv 0.$$

Moreover let us fix the saturation constraints

$$M_1 = 2, \quad M_2 = 3$$

and choose the initial datum $\mathbf{x}_0 = (5, -4, 11, 8, 4)$. The system evolution is shown in Figure 4.6 and Figure 4.7.

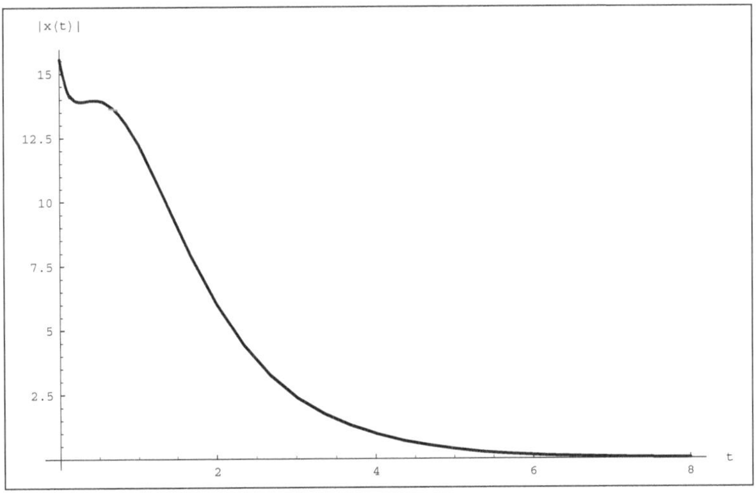

Fig. 4.6. Norm of the solution $\mathbf{x}(t)$ in the time interval (0,8)

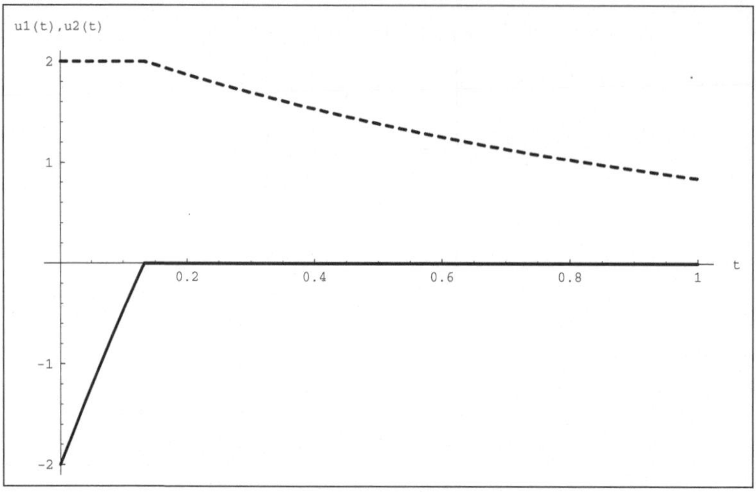

Fig. 4.7. Controls u_1 (dotted line) and u_2 (continuous line) in the time interval (0,1)

Example 4.2. Consider the following system $(\mathbf{A},\mathbf{B})$ with $n = 5,\ m = 3$:

$$\mathbf{A} = \begin{pmatrix} -3.25 & -3.41\overline{6} & -0.375 & 0 & -1.25 \\ -7.5 & -1 & 0 & 0 & -2.5 \\ 7.5 & 3 & 2 & 0 & 2.5 \\ 1.5 & 0.8 & 0 & 3 & 0.5 \\ 6.75 & -0.25 & 1.125 & 0 & 2.75 \end{pmatrix}$$

and

$$\mathbf{B} = \begin{pmatrix} 0.08\overline{3} & 0.\overline{5} & -0.625 \\ 4 & 0 & -2.5 \\ -3.\overline{6} & 0 & 2.5 \\ -0.8 & -0.4 & 0.5 \\ -2.25 & -0.\overline{3} & 2.875 \end{pmatrix}$$

The eigenvalues of $\mathbf{A}$ are $\nu_{1,2} = -15\text{i},\quad \nu_3 = 3,\quad \nu_4 = 2$ and $\nu_5 = 0.5$, so that the system has three unstable modes. Changing the coordinates the new expressions are

$$\overline{\mathbf{A}} = \begin{pmatrix} -1 & 5 & 0 & 0 & 0 \\ -5 & -1 & 0 & 0 & 0 \\ 0 & 0 & 3 & 0 & 0 \\ 0 & 0 & 0 & 2 & 0 \\ 0 & 0 & 0 & 0 & 0.5 \end{pmatrix}$$

and

$$\overline{\mathbf{B}} = \begin{pmatrix} -2 & 1.\overline{3} & 1 \\ 8 & 0 & -5 \\ 0.\overline{3} & 0 & 0 \\ 0 & -2 & 0 \\ 0 & 0 & -5 \end{pmatrix}.$$

Set the control variables

$$u_1(\mathbf{x}) = -3(2+\varepsilon)\mathbf{x}_3,$$

$$u_2(\mathbf{x}) = \frac{3+\varepsilon}{2}\mathbf{x}_4,$$

$$u_3(\mathbf{x}) = \frac{0.5+\varepsilon}{5}\mathbf{x}_5.$$

and the saturation constraints $M_1 = 2$, $M_2 = 4$, $M_3 = 1.8$. Figure 4.8 and Figure 4.9 illustrate the system evolution starting from $\mathbf{x}_0 = (10, -6.5, -0.2, 1.8, -5.7)$.

4.1.6.4 General Conditions for Set Invariance

In the previous sections conditions for the existence of SSIS linear feedbacks have been given for special classes of systems, namely systems with a single unstable mode and systems with the number of unstable real modes equal to the number of inputs.

Consider now an arbitrary MI system. The SSIS condition is in general not easy to be met by the system matrices. It is well known (see for instance [46], [47]) that the invariance of the control input sublevel-sets is equivalent to the existence of a solution $\mathbf{F}$ to the matricial equation

$$\mathbf{AK} + \mathbf{KBK} = \mathbf{FK};$$

moreover, in order to ensure positive invariance, the coefficients of $\mathbf{F}$ are required to satisfy some positivity constraints.

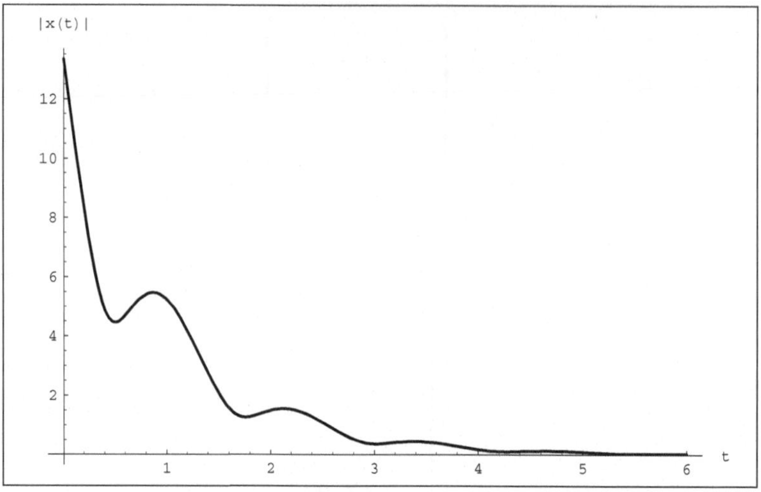

Fig. 4.8. Norm of the solution $\mathbf{x}(t)$ in the time interval (0,6)

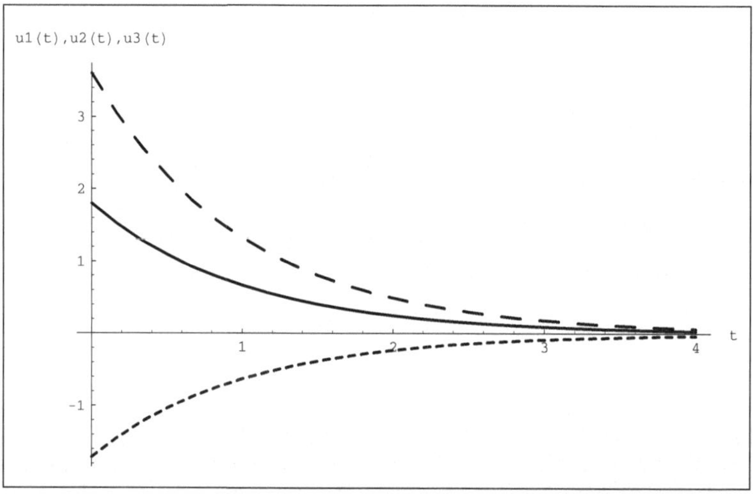

Fig. 4.9. Controls u_1 (continuous line), u_2 (dashed line) and u_3 (dotted line) in the time interval (0,4)

The main result presented in this section illustrates a different approach toward investigation on the existence of SSIS linear controls. For a multi input linear system having distinct real unstable modes, it is shown that, if the columns of the input matrix are sufficiently close to a set of eigenvectors corresponding to the positive eigenvalues of the state matrix **A**, a SSIS linear feedback can be found.

Assume the existence of m vectors $\mathbf{K}_j$, $j = 1,...,m$ with

$$\langle \mathbf{K}_j, \mathbf{x}(t)\rangle \langle \mathbf{K}_j, \dot{\mathbf{x}}(t)\rangle \leq 0. \tag{4.33}$$

The condition above is equivalent to the invariance of the set

$$\{\mathbf{x} \in \mathbb{R}^n : |\langle \mathbf{K}_j, \mathbf{x}\rangle| < M_j \ \forall\ j = 1,...,m\},$$

which guarantees the validity of SSIS property for the control

$$\mathbf{u}(t) = (\mathbf{K}_1^T \mathbf{x}(t), ..., \mathbf{K}_m^T \mathbf{x}(t)).$$

Suppose that for any $j = 1,...,m$ we have $\langle \mathbf{K}_j, \mathbf{x}(t)\rangle = M_j$; from (4.33), it follows that

$$\langle \mathbf{K}_j, \dot{\mathbf{x}}(t)\rangle \leq 0 \ \forall\ j = 1,...m. \tag{4.34}$$

Now

$$\langle \mathbf{K}_j, \dot{\mathbf{x}}(t)\rangle = \left\langle \mathbf{K}_j, \left(\mathbf{A} + \sum_{j=1}^{m} \mathbf{B}_j \mathbf{K}_j^T\right) \mathbf{x}(t) \right\rangle =$$
$$= \left\langle \mathbf{K}_j, \left(\mathbf{A} + \sum_{i \neq j} \mathbf{B}_i \mathbf{K}_i^T\right) \mathbf{x}(t) \right\rangle + \langle \mathbf{K}_j, \mathbf{B}_j \mathbf{K}_j^T \mathbf{x}(t)\rangle =$$
$$= \left\langle \left(\mathbf{A}^T + \sum_{i \neq j} \mathbf{K}_i \mathbf{B}_i^T\right) \mathbf{K}_j, \mathbf{x}(t) \right\rangle + M_j \langle \mathbf{K}_j, \mathbf{B}_j\rangle$$

From (4.34) follows the inequality

$$\left\langle \left(\mathbf{A}^T + \sum_{i \neq j} \mathbf{K}_i \mathbf{B}_i^T\right) \mathbf{K}_j, \mathbf{x}(t) \right\rangle \leq -M_j \mathbf{K}_j^T \mathbf{B}_j \tag{4.35}$$

whenever

$$\langle \mathbf{K}_j, \mathbf{x}(t)\rangle = M_j.$$

These conditions can be together fulfilled only if the spaces

$$\left\langle \left(\mathbf{A}^T + \sum_{i \neq j} \mathbf{K}_i \mathbf{B}_i^T\right) \mathbf{K}_j, \mathbf{x}(t) \right\rangle = 0$$

and

$$\langle \mathbf{K}_j, \mathbf{x}(t)\rangle = 0$$

are parallel; it turns out that $\mathbf{K}_j$ must be an eigenvector for the matrix $\mathbf{A}^T + \sum_{i \neq j} \mathbf{K}_i \mathbf{B}_i^T$. In particular for $j = 1,...,m$ the identity

$$\left(\mathbf{A}^T + \sum_{i \neq j} \mathbf{K}_i \mathbf{B}_i^T \right) \mathbf{K}_j = \mu_j \mathbf{K}_j \tag{4.36}$$

holds. From (4.35), it follows that the eigenvalues μ_j must satisfy

$$\mu_j \leq -\mathbf{K}_j^T \mathbf{B}_j. \tag{4.37}$$

The following statement summarizes the previous development.

Proposition 4.3. *Suppose that the system (4.28) admits an SSIS linear feedback controller, say* $\mathbf{u} = (u_1,...,u_m)$, *where* $u_j = \langle \mathbf{K}_j, \mathbf{x} \rangle$ *with* $\mathbf{K}_j \in \mathbb{R}^n$ *for* $j = 1,...,m$. *Then the vectors* $\mathbf{K}_j$ *satisfy (4.36) and (4.37).*

Remark 4.7. *If the column vectors* $\mathbf{B}_j$ *are a family of eigenvectors for the state matrix* $\mathbf{A}$, *it is very easy to verify that condition (4.36) is satisfied. As shown in Proposition 4.2, the SSIS feedback is constituted by a set of dual eigenvectors* $\mathbf{K}_j$, *i.e.* $\mathbf{A}^T \mathbf{K}_j = \lambda_j \mathbf{K}_j$. *Now, since* $\mathbf{B}_i \in ker(\mathbf{A} - \lambda_i \mathbf{I})$, $\mathbf{K}_j \in Im(\mathbf{A}^T - \lambda_i \mathbf{I})$ *and* $ker(\mathbf{A}) \perp Im(\mathbf{A}^T)$, *it follows that for any* $j = 1,...,m$

$$\langle \mathbf{B}_i, \mathbf{K}_j \rangle = \mathbf{B}_i^T \mathbf{K}_j = 0 \quad \forall i \neq j.$$

The existence of SSIS linear controls is ruled by nonlinear algebraic equations (4.36)-(4.37); nevertheless, the following invariance result holds.

Proposition 4.4. *The SSIS property is invariant under orthogonal coordinates transformations.*

Proof. Let $\mathbf{H}$ be a orthogonal matrix, i.e. $\mathbf{H}\mathbf{H}^T = \mathbf{H}^T \mathbf{H} = \mathbf{I}$, and let $\mathbf{A}$, $\{\mathbf{B}_i\}$, $\{\mathbf{K}_i\}$ and $\{\mu_i\}$ satisfy the SSIS conditions; by the change of coordinates associated to $\mathbf{H}$, the relations (4.36) and (4.37) can be rewritten as

$$\begin{aligned} &\left(\mathbf{H}^T \mathbf{A}^T \mathbf{H} + \textstyle\sum_{i \neq j} \mathbf{H}^T \mathbf{K}_i \mathbf{B}_i^T \mathbf{H}\right) \mathbf{H}^T \mathbf{K}_j - \mu_j \mathbf{H}^T \mathbf{K}_j = \\ &= \mathbf{H}^T \left(\left(\mathbf{A}^T + \textstyle\sum_{i \neq j} \mathbf{K}_i \mathbf{B}_i^T \right) \mathbf{K}_j - \mu_j \mathbf{K}_j \right) = 0 \end{aligned}$$

and

$$\mu_j \leq -\mathbf{K}_j^T \mathbf{H}\mathbf{H}^T \mathbf{B}_j = -\mathbf{K}_j^T \mathbf{B}_j. \qquad \square$$

Corollary 4.1. *If* $\mathbf{A}$, $\{\mathbf{B}_i\}$, $\{\mathbf{K}_i\}$ *and* $\{\mu_i\}$ *satisfy the SSIS conditions and an arbitrary change of coordinates is applied to the system by the transformation associated to the matrix* $\mathbf{H}$ *with* $\det(\mathbf{H}) \neq 0$, *then* $\{\mathbf{H}^{-1}\mathbf{K}_i\}$ *are SSIS linear feedbacks for the system* $(\mathbf{H}^{-1}\mathbf{A}\mathbf{H}, \{\mathbf{H}^T \mathbf{H}(\mathbf{H}^{-1}\mathbf{B}_i)\})$.

The next theorem states the local existence of SSIS linear feedbacks for an arbitrary system having distinct unstable real eigenvalues.

Theorem 4.5. *Suppose that* $\mathbf{A}$ *has* m *unstable real modes; due to Proposition 4.4 and Corollary 4.1, it can be assumed without loss of generality that* $\mathbf{A}$ *is given in Jordan canonical form. Let us denote with* $\overline{\mathbf{B}}_i$, $i = 1,...,m$, *a family of independent eigenvectors corresponding to the positive distinct real eigenvalues* $\nu_1,...,\nu_m$. *There exist* $\rho_i > 0$ *such that for* $\mathbf{B}_i \in \mathscr{C}_{\rho_i}(\overline{\mathbf{B}}_i) := \{\mathbf{V} \in \mathbb{R}^n : ||\mathbf{V} - \overline{\mathbf{B}}_i|| < \rho_i\}$, *the system (4.28) admits SSIS linear feedbacks.*

Proof. The existence of SSIS controls will be proved applying the general implicit function theorem. The following notations are employed: $\mathbf{B}_i = \{b_j^{(i)}\}_{j=1,...,n}$, $\mathbf{K}_i = \{k_j^{(i)}\}_{j=1,...,n}$, $k_i^{(i)} = \chi_i$ and $\mathbf{A} = \mathrm{diag}(\mathbf{A}_u, \mathbf{A}_s)$ with $\mathbf{A}_u = \mathrm{diag}(\nu_1,...,\nu_m)$ and $\mathbf{A}_s = \{a_{rs}\}_{r,s=m+1,...,n}$. The SSIS conditions (4.36) can be rewritten as the following set of relations for $i = 1,...,m$:

$$
\begin{cases}
\Phi_r^{(i)} = \left(\nu_r - \mu_i + \sum_{j\neq i} k_r^{(j)} b_r^{(j)}\right) k_r^{(i)} \\
\quad + \sum_{s=1,\, s\neq r}^{n} \left(\sum_{j\neq i} k_r^{(j)} b_s^{(j)}\right) k_s^{(i)} = 0 & r = 1,...,m, \\
\Phi_r^{(i)} = \sum_{1\leq s<r} \left(\sum_{j\neq i} k_r^{(j)} b_s^{(j)}\right) k_s^{(i)} + \left(a_{rr} - \mu_i + \sum_{j\neq i} k_r^{(j)} b_r^{(j)}\right) k_r^{(i)} + \\
\quad + \sum_{r<s\leq n} \left(a_{rs} + \sum_{j\neq i} k_r^{(j)} b_s^{(j)}\right) k_s^{(i)} = 0 & r = m+1,...,n.
\end{cases}
$$

Let us consider the $n \cdot m$ dimensional vector

$$\mathbf{y} = (k_2^{(1)}, k_3^{(1)}, ..., \mu_1, k_1^{(2)}, k_3^{(2)}, ..., k_{j-1}^{(j)}, k_{j+1}^{j}, ..., k_n^{(m)}, \mu_m)$$

given by μ_j and all entries of K_j except $k_j^{(j)}$. Let us consider in addiction the mapping

$$\mathscr{F} : \mathbb{R}^{n\cdot m} \times \mathbb{R}^{n\cdot m} \to \mathbb{R}^{n\cdot m}$$

given by

$$\mathscr{F}(\mathbf{B}_1, ..., \mathbf{B}_m, \mathbf{y}) = (\Phi_1^{(1)}, \Phi_2^{(1)}, ..., \Phi_{n-1}^{(m)}, \Phi_n^{(m)})$$

The SSIS constraint (4.36) is equivalent to the equation $\mathscr{F}(\mathbf{B}_1, ..., \mathbf{B}_m, \mathbf{y}) = 0$. Referring to Remark 4.7, it can be seen that, setting $\overline{\mathbf{B}}_i = \lambda_i \mathbf{e}_i, \overline{\mu}_i = \nu_i$ and $\overline{\mathbf{K}}_i = h_i \mathbf{e}_i$, it follows $\mathscr{F}(\overline{\mathbf{B}}_1, ..., \overline{\mathbf{B}}_m, \overline{\mathbf{y}}) = 0$. In order to apply the implicit function theorem, the following non-singularity condition has to be satisfied

$$\det \frac{\partial \mathscr{F}}{\partial \mathbf{y}}\Big|_{(\overline{\mathbf{B}}_1, ..., \overline{\mathbf{B}}_m, \overline{\mathbf{y}})} \neq 0$$

The following quantities have to be taken into account:

$$\frac{\partial \Phi_r^{(i)}}{\partial k_j^{(i)}} = \begin{cases} \delta_{jr}\left(\nu_r - \mu_i + \sum_{s\neq i} k_r^{(s)} b_r^{(s)}\right) \\ +(1-\delta_{jr})\sum_{s\neq i} k_r^{(s)} b_j^{(s)} & r \leq m \\ \\ \delta_{jr}\left(a_{rr} - \mu_i + \sum_{s\neq i} k_r^{(s)} b_r^{(s)}\right) \\ +(1-\delta_{jr})\sum_{s\neq i} k_r^{(s)} b_j^{(s)} & r > m, \quad j \leq r \\ \\ a_{rj} + \sum_{s\neq i} k_r^{(s)} b_j^{(s)} & r > m, \quad j > r \end{cases}$$

$$\frac{\partial \Phi_r^{(i)}}{\partial k_j^{(s)}} = \delta_{rj} \sum_{h\neq r} b_h^{(s)} k_h^{(i)} \qquad s \neq i$$

$$\frac{\partial \Phi_r^{(i)}}{\partial \mu_i} = -k_r^{(i)}$$

$$\frac{\partial \Phi_r^{(i)}}{\partial \mu_s} = 0 \qquad s \neq i$$

Computing the above derivatives and performing the evaluation in $(\overline{\mathbf{B}}_1,...,\overline{\mathbf{B}}_m,\overline{\mathbf{y}})$ a block diagonal matrix is obtained:

$$\frac{\partial \mathscr{F}}{\partial \mathbf{y}}_{|(\overline{\mathbf{B}}_1,...,\overline{\mathbf{B}}_m,\overline{\mathbf{y}})} = \mathrm{Diag}(\mathscr{D}_1,...,\mathscr{D}_m);$$

each block $\mathscr{D}_j$ is a $n \times n$ matrix given by

$$\mathscr{D}_j = \begin{pmatrix} \mathrm{diag}(\nu_s - \nu_j + h_s\lambda_s)_{s=1}^{j-1} & \mathbf{0}_{j-1\times m-j} & \mathbf{0}_{j-1\times n-m} & 0 \\ \mathbf{0}_{1\times j-1} & \mathbf{0}_{1\times m-j} & \mathbf{0}_{1\times n-m} & -h_j \\ \mathbf{0}_{m-j\times j-1} & \mathrm{diag}(\nu_s - \nu_j + h_s\lambda_s)_{s=j+1}^{m} & \mathbf{0}_{m-j\times n-m} & 0 \\ \mathbf{0}_{n-m\times j-1} & \mathbf{0}_{n-m\times m-j} & \mathbf{A}_s - \nu_j \mathbf{I}_{n-m} & 0 \end{pmatrix}$$

Moreover, the block $\mathscr{D}_j$ is itself a block-diagonal matrix up to a row ordering rearrangement. Now

$$\det \frac{\partial \mathscr{F}}{\partial \mathbf{y}}_{|(\overline{\mathbf{B}}_1,...,\overline{\mathbf{B}}_m,\overline{\mathbf{y}})} = \prod_{j=1}^{m} \det \mathscr{D}_j.$$

Let us fix $\varepsilon_j > 0$ and choose h_j such that condition (4.37) holds in the quantitative form

$$-\overline{\mathbf{K}}_j \overline{\mathbf{B}}_j - \mu_j = \varepsilon_j > 0, \tag{4.38}$$

that is $h_j = -(\nu_j + \varepsilon_j)/\lambda_j$. Now $\nu_s - \nu_j + h_s\lambda_s = \varepsilon_s - \nu_j = \varepsilon_s + |\nu_j| > 0$. It follows that

$$\det \mathscr{D}_j = (-1)^{n-j+1} h_j \cdot \det(\mathbf{A}_s - \nu_j \mathbf{I}_{n-m}) \prod_{s\neq j, s=1}^{m} (\varepsilon_s + |\nu_j|) \neq 0.$$

By the implicit function theorem there exists a neighborood J of $(\overline{\mathbf{B}}_1,...,\overline{\mathbf{B}}_m)$ such that (4.36) is verified with $\mathbf{y}=\mathbf{y}(\overline{\mathbf{B}}_1,...,\overline{\mathbf{B}}_m)$. Moreover, as a consequence of sign permanency, there exist $J_1 \subseteq J$ such that (4.37) is satisfied too. This concludes the proof. □

The global existence of SSIS controls can be investigated following a criterion introduced in [48]. Suppose that a real continuous positive function $z:[0,\infty)\rightarrow[0,\infty)$ with $\int_1^{\infty}\frac{1}{z(s)}ds=\infty$ can be found such that

$$||\mathscr{F}(\mathbf{B}_1,...,\mathbf{B}_m,\mathbf{y})||\cdot||(\mathscr{F}(\mathbf{B}_1,...,\mathbf{B}_m,\mathbf{y}))^{-1}||<z(||\mathbf{y}||)\quad \text{for } \mathbf{B}_i\in J_i \text{ and } \mathbf{y}\in J_{\mathbf{y}},$$

where $J_i\subset\mathbb{R}^n$ and $J_{\mathbf{y}}\subset\mathbb{R}^{n\cdot m}$. Then there exists an implicit function $\mathbf{y}=\mathbf{y}(\mathbf{B}_1,...,\mathbf{B}_m)$ satisfying

$$\mathscr{F}(\mathbf{B}_1,...,\mathbf{B}_m,\mathbf{y}(\mathbf{B}_1,...,\mathbf{B}_m))=0$$

for any $(\mathbf{B}_1,...,\mathbf{B}_m)\in J_1\times J_2\cdots\times J_m$.

Remark 4.8. *As in the SISO case (see [49]), the widest region of attraction associated to SSIS controls is the open set* $E_{\mathbf{M}}=\{\mathbf{x}\in\mathbb{R}^n : |\langle\mathbf{K}_i,\mathbf{x}\rangle|<M_i,\ i=1,...,m\}$ *where* $\mathbf{K}_i$ *verifies (4.36) and where condition (4.37) is satisfied with equality for any* $i=1,...,m$.

4.2 Nonlinear Robust Controller Design via Sliding Modes for Continuous-Time Multi Input Plants

This section is devoted to the description of some results about the stabilization of multi-input plants by sliding modes [50], [51]. As well known, the main advantage of the sliding mode control technique is robustness, i.e. invariance with respect to bounded matched disturbances. It will be proved that robust stabilization of linear plants can be achieved by means of a time-varying state feedback controller, derived imposing the achievement of a sliding motion onto a suitable time-varying sliding surface [42] [43].The presented results can be found in [52] [53] [41].

Consider the following continuous-time, time invariant, controllable, uncertain Multi-Input plant $S\overset{\text{def}}{=}\{\hat{\mathbf{A}},\hat{\mathbf{B}}\}$:

$$\dot{\hat{\mathbf{x}}}=\hat{\mathbf{A}}\hat{\mathbf{x}}+\hat{\mathbf{B}}(\mathbf{I}+\Delta\hat{\mathbf{B}})(\hat{\mathbf{u}}+\hat{\mathbf{d}}(t)) \tag{4.39}$$

where: $\hat{\mathbf{x}}=[\hat{x}_1\cdots\hat{x}_n]^T\in\mathbb{R}^n$ is the state vector (assumed available for measurement), $\hat{\mathbf{u}}\in\mathbb{R}^m$ is the control input, with $n>m$, and $\hat{\mathbf{A}}\in\mathbb{R}^{n\times n}$, $\hat{\mathbf{B}}\in\mathbb{R}^{n\times m}$ are the state matrix and the input matrix respectively. The matched uncertain terms $\hat{\mathbf{d}}(t)\in\mathbb{R}^m$ and $\Delta\hat{\mathbf{B}}\in\mathbb{R}^{m\times m}$ represents plant uncertainties and/or external disturbances affecting the system, and $\mathbf{I}$ is the m dimensional identity matrix. Under the controllability hypothesis, there exist two smooth changes of coordinates: $\mathbf{x} = \mathbf{T_x}\hat{\mathbf{x}}$, $\mathbf{u} = \mathbf{T_u}\hat{\mathbf{u}}$ such that state and the input matrices of plant (4.39) have a canonical structure depending on the controllability indices of the system ρ_j, $j = 1 \dots m$ [54] [55]. Rearranging rows and columns [55], plant (4.39) can be further transformed into:

$$\dot{\mathbf{x}} = \mathbf{A}\mathbf{x} + \mathbf{B}(\mathbf{I} + \Delta\mathbf{B})(\mathbf{u} + \mathbf{d}(t)) \tag{4.40}$$

with $\mathbf{d}(t) = \mathbf{T_u}\hat{\mathbf{d}}(t)$ and $\mathbf{B} = \begin{bmatrix} \mathbf{0}_{(n-m)\times m} & \mathbf{I} \end{bmatrix}^T$, and: $\mathbf{A} = \begin{bmatrix} \mathbf{A}_{1,1} & \mathbf{A}_{1,2} \\ \mathbf{A}_{2,1} & \mathbf{A}_{2,2} \end{bmatrix}$, with $\mathbf{A}_{1,1} \in \mathbb{R}^{(n-m)\times(n-m)}$, $\mathbf{A}_{1,2} \in \mathbb{R}^{(n-m)\times m}$, $\mathbf{A}_{2,1} \in \mathbb{R}^{m\times(n-m)}$, $\mathbf{A}_{2,2} \in \mathbb{R}^{m\times m}$, where the block matrix $\mathbf{A}_1 = \begin{bmatrix} \mathbf{A}_{1,1} & \mathbf{A}_{1,2} \end{bmatrix}$ has elements $\{\xi_{i,j}\}$, $i = 1, \dots, n-m$, $j = 1, \dots, n$ which are either 1 or 0, and $\mathbf{A}_2 = \begin{bmatrix} \mathbf{A}_{2,1} & \mathbf{A}_{2,2} \end{bmatrix} = \begin{bmatrix} a_{1,1} & a_{1,2} & \dots & a_{1,n} \\ & & \dots & \\ a_{m,1} & a_{m,2} & \dots & a_{m,n} \end{bmatrix}$.

The plant is preceded by a saturating device $\mathbf{u} = f(\mathbf{v})$ with known threshold $M > 0$, i.e. for the i-th component:

$$u_i = \text{sat}_M(v_i) \quad i = 1, \dots, m. \tag{4.41}$$

Assumption 4.1. *The uncertain terms $\mathbf{d}(t)$ and $\Delta\mathbf{B}$ are bounded by known constants, i.e. $||\mathbf{d}(t)|| \le \bar{\rho}$, $||\Delta\mathbf{B}|| \le \delta$.*

Exploiting the presence of the saturation, system equations can be rearranged as follows

$$\dot{\mathbf{x}} = \mathbf{A}\mathbf{x} + \mathbf{B}(\mathbf{u} + \mathbf{d}_1(t)) \tag{4.42}$$

with $\mathbf{d}_1(t) = \mathbf{d}(t) + \Delta\mathbf{B}(\mathbf{u} + \mathbf{d}(t))$, $||\mathbf{d}_1(t)|| \le \bar{\rho} + \delta(M + \bar{\rho})$. Consider a matrix $\mathbf{C} \in \mathbb{R}^{m\times n}$ of the form

$$\mathbf{C} = \begin{bmatrix} \mathbf{C}_1 & \mathbf{C}_2 \end{bmatrix} = \begin{bmatrix} c_{1,1} & \dots & c_{1,n-m} & \varepsilon_1 & 0 & \dots \\ & \dots & & & & \dots \\ c_{m,1} & \dots & c_{m,n-m} & 0 & 0 & \varepsilon_m \end{bmatrix} \tag{4.43}$$

with $\mathbf{C}_1 \in \mathbb{R}^{(n-m)\times m}$, $\mathbf{C}_2 \in \mathbb{R}^{m\times m}$, $c_{i,j}, \varepsilon_i \in \mathbb{R}$, $i = 1, \dots, m$, $j = 1, \dots, n-m$. Coefficients appearing in the $\mathbf{C}_1$ matrix can be designed such that, when a sliding motion [50] is achieved on the following sliding surface:

$$\hat{\mathbf{s}}(\mathbf{x}) = \mathbf{C}\mathbf{x} = \mathbf{0} \tag{4.44}$$

the corresponding reduced order system has assigned stable eigenvalues, and, as a consequence, system (4.42) is stable. Indeed, it has been shown [56] [57] that the design problem of $\mathbf{C}$ is equivalent to an eigenvalue assignment problem which, under the controllability hypothesis, can always be solved. In the absence of the saturating device (4.41), i.e. if the control input $\mathbf{u}$ could be directly manipulated, the following control law, obtained by imposing the inequality $\hat{\mathbf{s}}^T(\mathbf{x})\dot{\hat{\mathbf{s}}}(\mathbf{x}) < 0$,

$$\mathbf{u}_c = -(\mathbf{C}\mathbf{B})^{-1}\mathbf{C}\mathbf{A}\mathbf{x} - \bar{\rho}\,sign(\mathbf{B^T}\mathbf{C^T}\hat{\mathbf{s}}(\mathbf{x})) \tag{4.45}$$

would ensure the achievement of a sliding motion on (4.44), hence plant stabilization[1]. Since the input $\mathbf{v}$ only can be manipulated, the control problem addressed in this paper consists in finding a feedback controller $\mathbf{v}$ guaranteeing the robust stabilization of the system (4.39) in the presence of a saturating nonlinearity in the actuator.

4.2.1 A Time Varying Sliding Surface

Define the matrix $\mathbf{D} \in \mathbb{R}^{m\times n}$ as

$$\mathbf{D} = \begin{bmatrix} \mathbf{D}_1 & \mathbf{0} \end{bmatrix} = \begin{bmatrix} d_{1,1} & \dots & d_{1,n-m} & 0 & 0 \\ & \dots & & \dots & \\ d_{m,1} & \dots & d_{m,n-m} & 0 & 0 \end{bmatrix} \tag{4.46}$$

and $\bar{\mathbf{C}}(t) \stackrel{\text{def}}{=} (\mathbf{C} + \mathbf{D}e^{(-\bar{\lambda}t)}) = [\bar{c}_{i,j}]$. Consider the following time-varying sliding surface, with $\bar{\lambda} > 0$:

$$\mathbf{s}(\mathbf{x}(t),\mathbf{x}(0),t) = \bar{\mathbf{C}}(t)\left[\mathbf{x}(t) - \mathbf{x}(0)e^{(-\bar{\lambda}t)}\right] = 0 \tag{4.47}$$

$$= \begin{bmatrix} \bar{c}_{1,1}\dots\bar{c}_{1,n-m} & \\ \vdots & diag\{\varepsilon_1,\dots,\varepsilon_m\} \\ \bar{c}_{m,1}\dots\bar{c}_{m,n-m} & \end{bmatrix} \left[\mathbf{x}(t) - \mathbf{x}(0)e^{(-\bar{\lambda}t)}\right]$$

where $\bar{c}_{i,j} = (c_{i,j} + d_{i,j}e^{(-\bar{\lambda}t)})$. It can be proved (see Appendix) that, for any choice of $d_{i,j} \lesseqgtr 0$, $i = 1,m$, $j = 1,\dots,n-m$, constraining the system to the surface $\mathbf{s}(\mathbf{x}(t),\mathbf{x}(0),t) = 0$ implies plant asymptotical stabilization. Moreover, since $\mathbf{s}(\mathbf{x}(0),\mathbf{x}(0),0) = 0$, the surface (4.47) is such that no reaching phase exists [42] [43], as the plant state is on the hyperplane from the very beginning. What motivates the introduction of the vanishing term $\mathbf{D}e^{(-\bar{\lambda}t)}\mathbf{x}$ with respect to standard surfaces is the need of modulating the control input in order to cope with the saturation limitation. Roughly speaking, we are aiming at constraining the system on a sliding surface which, besides being asymptotically stabilizing, has a tunable part such that the control input is able to constrain the plant state on the sliding hyperplane without violating the saturation bounds. The following section is therefore devoted to show that the coefficients of the $\mathbf{D}$ matrix can always be found as to satisfy the saturation limits, still preserving the persistence of the sliding motion.

For the surface (4.47), the control input ensuring the achievement of a finite-time sliding motion is, similarly to (4.45):

[1] The $sign(\cdot)$ symbol above denotes a vector whose i-th entry contains the sign of the i-th component of the vector $\mathbf{B}^{\mathrm{T}}\mathbf{C}^{\mathrm{T}}\hat{\mathbf{s}}(\mathbf{x})$.

$$\mathbf{v} = -\left(\bar{\mathbf{C}}(t)\mathbf{B}\right)^{-1}\left[(\bar{\mathbf{C}}(t)\mathbf{A} - \bar{\lambda}\mathbf{D}e^{(-\bar{\lambda}t)})\mathbf{x} + \varphi(\mathbf{x}(0),t)\right] \\ - \bar{\rho}\,sign(\mathbf{B}^T\bar{\mathbf{C}}(t)^T\mathbf{s}(\mathbf{x}(t),\mathbf{x}(0),t)) \tag{4.48}$$

where: $\varphi(\mathbf{x}(0),t) = \bar{\lambda}(\mathbf{C}+2\mathbf{D}e^{(-\bar{\lambda}t)})\mathbf{x}(0)e^{(-\bar{\lambda}t)}$. With some abuse of notation, the sign function used in (4.48) denotes a vector containing signs of the component of argument entries. As already discussed, for the asymptotically stabilizing surface (4.47) no reaching phase exists. Hence, the plant is in sliding motion from $t = 0$, and the dynamics of the state variables are governed by sliding mode. It follows that state trajectories are bounded.

Remark 4.9. *For any initial condition* $\mathbf{x}(0)$, *suitable functions bounding state trajectories can be determined from* $\mathbf{x}(0)$ *itself and the sliding mode dynamics imposed by the control law (4.48). It follows that, for any* $\mathbf{x}(0)$ *there exist a function* $\Delta_F^{(max)} \in \mathbb{R}^+$ *such that:*

$$||\mathbf{x}(t)|| \leq \Delta_F^{(max)}, \quad \forall t. \tag{4.49}$$

The dependence of such function on the parameters used for tuning the control algorithm will be addressed later, after a deep description of the control law.

Assumption 4.2. *It is assumed that the saturation threshold* M *satisfies* $M - \bar{\rho} - \delta(M+\bar{\rho}) \overset{\text{def}}{=} \bar{M} > 0$

4.2.2 The Control Law

The constraint induced by saturation (4.41) requires:

$$|v_i| \leq M, \quad i = 1,\ldots,m \tag{4.50}$$

Due to the form of the matrices $\mathbf{B}$, $\bar{\mathbf{C}}(t)$ (4.43), (4.46), one has $(\bar{\mathbf{C}}(t)\mathbf{B})^{-1} = diag\left\{\frac{1}{\varepsilon_1},\ldots,\frac{1}{\varepsilon_m}\right\}$. From expression (4.48), it follows:

$$v_i = -\frac{1}{\varepsilon_i}\left[\bar{\lambda}\sum_{i=1}^{n}\left(c_{i,j} + 2d_{i,j}e^{(-\bar{\lambda}t)}\right)x_j(0)e^{(-\bar{\lambda}t)} + \varepsilon_i\sum_{i=1}^{n}a_{i,j}x_j \right. \\ \left. + \sum_{j=1}^{n-m}\bar{c}_{i,j}\sum_{k=1}^{n}\xi_{j,k}x_k - \bar{\lambda}e^{(-\bar{\lambda}t)}\sum_{j=1}^{n-m}d_{i,j}x_j\right] - \bar{\rho}\,sign(\bar{s}_i(\mathbf{x})) \leq M \tag{4.51}$$

denoting the $i-$th component of $\mathbf{B}^T\ \bar{\mathbf{C}}(t)^T\ \mathbf{s}(\mathbf{x}(t),\mathbf{x}(0),t)$ as $\bar{s}_i(\mathbf{x})$ with a slight abuse of notation. Taking the worst case, the condition (4.50) for the $i-$th component can be rewritten as

$$\left| \bar{\lambda} \sum_{i=1}^{n} \left(c_{i,j} + 2d_{i,j} e^{(-\bar{\lambda} t)} \right) x_j(0) e^{(-\bar{\lambda} t)} + \varepsilon_i \sum_{i=1}^{n} a_{i,j} x_j \right.$$
$$\left. + \sum_{j=1}^{n-m} \bar{c}_{i,j} \sum_{k=1}^{n} \xi_{j,k} x_k - \bar{\lambda} e^{(-\bar{\lambda} t)} \sum_{j=1}^{n-m} d_{i,j} x_j \right| \le \bar{M} |\varepsilon_i| \tag{4.52}$$

Taking again the worst case, one has, for $i = 1, \dots, m$:

$$\sum_{i=1}^{n} \left[\bar{\lambda} \left(|c_{i,j}| + 2|d_{i,j}| \right) \frac{|x_j(0)|}{\Delta_F^{(max)}} + |\varepsilon_i| |a_{i,j}| \right]$$
$$+ \sum_{j=1}^{n-m} \left[n \left(|c_{i,j}| + |d_{i,j}| \right) + \bar{\lambda} |d_{i,j}| \right] \le \frac{\bar{M}}{m \Delta_F^{(max)}} |\varepsilon_i| \tag{4.53}$$

Theorem 4.6. *It is given the uncertain system (4.39) preceded by the saturating device (4.41), under Assumptions 4.1, 4.2. For initial conditions belonging to a suitable neighborhood of the origin, coefficients $d_{i,j}$, ε_i, $i = 1, \dots, m$, $j = 1, \dots, n-m$ and a $\bar{\lambda} > 0$ can be chosen such that the feedback controller (4.48) guarantees the robust asymptotical stabilization of the plant.*

Proof. The proof is constructive. Condition (4.53) corresponds to:

$$\sum_{i=1}^{n} \left[\bar{\lambda} \left(|c_{i,j}| + 2|d_{i,j}| \right) \frac{|x_j(0)|}{\Delta_F^{(max)}} + |\varepsilon_i| |a_{i,j}| \right]$$
$$+ \sum_{j=1}^{n-m} \left[n \left(|c_{i,j}| + |d_{i,j}| \right) + \bar{\lambda} |d_{i,j}| \right] \le M_1 |\varepsilon_i| \sum_{j=1}^{n-m+1} m_j^{-1} \tag{4.54}$$

with $M_1 \overset{\text{def}}{=} \dfrac{\bar{M}}{m \Delta_F^{(max)}}$, and where m_j, $j = 1, \dots, n-m+1$ are such that

$$\sum_{j=1}^{n-m+1} m_j^{-1} \le 1. \tag{4.55}$$

In the following, the imposition of condition (4.54) will be performed taking suitable $n-m+1$ "portions" of (4.54) itself, and designing control coefficients $d_{i,j}$, m_i, ε_i, $\bar{\lambda}$ involved in each derived inequality in order to ensure the simultaneous fulfillment of all of them.

Step 1. To start, consider a portion of condition (4.54) made of some terms corresponding to $j = 1$ in summations:

$$\bar{\lambda} \left(|c_{i,1}| + 2|d_{i,1}| \right) \frac{|x_1(0)|}{\Delta_F^{(max)}} + |\varepsilon_i| |a_{i,1}| + \bar{\lambda} |d_{i,1}| \le M_1 |\varepsilon_i| m_1^{-1}$$

Note that the terms $n\left(|c_{i,1}|+|d_{i,1}|\right)+|\varepsilon_i||a_{i,1}|\theta_2$ have not been taken, since they will be considered in the next step. Define $0 \le \theta_1 \le 1$ satisfying $\theta_1 |a_{i,1}| m_1 < M_1$, and choose $d_{i,1}$ such that:

$$\bar{\lambda}|d_{i,1}| < \frac{|\varepsilon_i|\left(\dfrac{M_1}{m_1} - \theta_1 |a_{i,1}|\right) - \bar{\lambda}|c_{i,1}|\dfrac{|x_1(0)|}{\Delta_F^{(max)}}}{(1+2\dfrac{|x_1(0)|}{\Delta_F^{(max)}})} \tag{4.56}$$

The previous inequality requires, in turn, that one sets $|\varepsilon_i| > \dfrac{\bar{\lambda}|c_{i,1}|\dfrac{|x_1(0)|}{\Delta_F^{(max)}}}{\dfrac{M_1}{m_1} - \theta_1 |a_{i,1}|} \stackrel{\text{def}}{=} Q_{i,1}$.

After these choices, and defining $\theta_2 \stackrel{\text{def}}{=} (1-\theta_1)$ and $\left(1 - \dfrac{|x_i(0)|}{\Delta_F^{(max)}}\right) \stackrel{\text{def}}{=} g_i(\mathbf{x}(0)) > 0$, $i = 1,\dots,m$, what remains of inequality (4.54) is:

$$\begin{aligned}
&\theta_2|\varepsilon_i||a_{i,1}| + n|c_{i,1}|g_1(\mathbf{x}(0)) + \frac{n|\varepsilon_i|}{\bar{\lambda}}\left(\frac{M_1}{m_1} - \theta_1|a_{i,1}|\right) + |\varepsilon_i||a_{i,2}| \\
&+\bar{\lambda}\left(|c_{1,2}|+2|d_{1,2}|\right)\frac{|x_2(0)|}{\Delta_F^{(max)}} + \bar{\lambda}|d_{i,2}| + n|c_{i,2}| + n|d_{i,2}| \\
&+\sum_{j=3}^{n}\left[\bar{\lambda}\left(|c_{i,j}|+2|d_{i,j}|\right)\frac{|x_j(0)|}{\Delta_F^{(max)}} + |\varepsilon_i||a_{i,j}|\right] + \sum_{j=3}^{n-m}[n\left(|c_{i,j}|\right. \\
&\left.+|d_{i,j}|\right) + \bar{\lambda}|d_{i,j}|] \le M_1|\varepsilon_i| \sum_{j=2}^{n-m+1} m_j^{-1}.
\end{aligned}$$

Step 2. Following the lines of the previous step, consider $j = 2$ (j even), define $\xi_i(\mathbf{x}(0)) \stackrel{\text{def}}{=} \dfrac{|x_i(0)|}{\Delta_F^{(max)}}$ and choose $d_{i,2}$ such that:

$$\bar{\lambda}|d_{i,2}| < \frac{|\varepsilon_i|\left(M_1\mu_2 - \nu_2\right) - n|c_{i,1}|g_1(\mathbf{x}(0)) - \bar{\lambda}|c_{i,2}|\xi_2(\mathbf{x}(0))}{1+2\xi_2(\mathbf{x}(0))} \tag{4.57}$$

with $\mu_2 \stackrel{\text{def}}{=} 1/m_2 - n/(\bar{\lambda}m_1)$ and $\nu_2 \stackrel{\text{def}}{=} |a_{i,2}| + \theta_2|a_{i,1}| - \dfrac{n}{\bar{\lambda}^k}\theta_1|a_{i,1}|$. As in the previous step, consistency of condition (4.57) is always ensured if one sets:

$$|\varepsilon_i| > \frac{\bar{\lambda}|c_{i,2}|\xi_2(\mathbf{x}(0)) + n|c_{i,1}|g_1(\mathbf{x}(0))}{(M_1\mu_2 - \nu_2)} \stackrel{\text{def}}{=} Q_{i,2}$$

provided that $M_1\mu_2 - \nu_2 > 0$ (the numerator of $Q_{i,2}$ is always positive). A possible strategy to ensure this latter condition is to select $\mu_2 > 0$ and $\nu_2 < -\bar{\lambda}$, corresponding to:

$$\bar{\lambda} > \frac{nm_2}{m_1} \stackrel{\text{def}}{=} \bar{\lambda}_1; \quad \bar{\lambda} < \bar{\lambda}_2. \tag{4.58}$$

where $\bar{\lambda}_2$ is the positive solution of the equation: $\bar{\lambda}^2 + (|a_{i,2}| + \theta_2|a_{i,1}|)\bar{\lambda} - n\theta_1|a_{i,1}| = 0$. Again, what remains of (4.54) to be fulfilled in the previous steps is:

$$\begin{aligned}
&n|c_{i,2}|g_2(\mathbf{x}(0)) - n^2\frac{|c_{i,1}|}{\bar{\lambda}}g_1(\mathbf{x}(0)) + \frac{n|\varepsilon_i|}{\bar{\lambda}}(M_1\mu_2 - \nu_2) \\
&+\bar{\lambda}\left(|c_{i,3}| + 2|d_{i,3}|\right)\frac{|x_3(0)|}{\Delta_F^{(max)}} + |\varepsilon_i||a_{i,3}| + \bar{\lambda}|d_{i,3}| + n|c_{i,3}| \\
&+n|d_{i,3}| + \sum_{j=4}^{n}\left[\bar{\lambda}\left(|c_{i,j}| + 2|d_{i,j}|\right)\frac{|x_j(0)|}{\Delta_F^{(max)}} + |\varepsilon_i||a_{i,j}|\right] \\
&+\sum_{j=4}^{n-m}\left[n\left(|c_{i,j}| + n|d_{i,j}|\right) + \bar{\lambda}|d_{i,j}|\right] \le M_1|\varepsilon_i|\sum_{j=3}^{n-m+1} m_j^{-1}
\end{aligned} \tag{4.59}$$

Step 3. The case $j = 3$ (j odd) can be treated similarly, though some different choices are needed. As before, define

$$\mu_3 \stackrel{\text{def}}{=} \frac{1}{m_3} - \frac{n}{\bar{\lambda}^k}\mu_2; \quad \nu_3 \stackrel{\text{def}}{=} -\frac{n\nu_2}{\bar{\lambda}} \tag{4.60}$$

and choose $d_{i,3}$ such that:

$$\begin{aligned}
\bar{\lambda}|d_{i,3}| < &\frac{1}{(1 + 2\xi_3(\mathbf{x}(0)))}\{|\varepsilon_i|\left(M_1\mu_3 - \nu_3\right) - \bar{\lambda}|c_{i,3}|\xi_3(\mathbf{x}(0)) \\
&+\frac{n^2|c_{i,1}|}{\bar{\lambda}}g_1(\mathbf{x}(0)) - n|c_{i,2}|g_2(\mathbf{x}(0))\}.
\end{aligned} \tag{4.61}$$

The previous inequality requires to set:

$$\begin{aligned}
|\varepsilon_i| > &\frac{-n^2\frac{|c_{i,1}|}{\bar{\lambda}}g_1(\mathbf{x}(0)) + n|c_{i,2}|g_2(\mathbf{x}(0)) + \bar{\lambda}|c_{i,3}|\xi_i(\mathbf{x}(0))}{(M_1\mu_3 - \nu_3)} \\
&\stackrel{\text{def}}{=} Q_{i,3}
\end{aligned} \tag{4.62}$$

and $M_1\mu_3 - \nu_3 > 0$, or equivalently $\frac{M_1}{m_3} > \frac{n}{\bar{\lambda}^k}(\frac{M_1}{m_2} + \frac{\theta_1|a_{i,1}|}{\bar{\lambda}})$ i.e. $\bar{\lambda} > \bar{\lambda}_3$ for a suitable $\bar{\lambda}_3 > 0$.

Step j. The above procedure can be generalized for any $j = r \leq n - m$, but the cases j even and j odd need to be differentiated. Define:

$$\mu_1 = \frac{1}{m_1}; \quad \mu_r \stackrel{\text{def}}{=} \frac{1}{m_r} - \frac{n}{\bar{\lambda}^k}\mu_{r-1}; \quad r = 2 \ldots n-m;$$
$$\nu_1 \stackrel{\text{def}}{=} \theta_1 |a_{i,1}|; \quad \nu_2 \stackrel{\text{def}}{=} |a_{i,2}| + \theta_2 |a_{i,1}| - \frac{n}{\bar{\lambda}^k}\nu_1 \tag{4.63}$$
$$\nu_r \stackrel{\text{def}}{=} \begin{cases} -\dfrac{n\nu_{r-1}}{\bar{\lambda}}, & 2 < r < n-m, \quad r\ odd; \\ \sum_{\ell=r-1}^{r} |a_{i,\ell}| + \dfrac{n^2 \nu_{r-2}}{\bar{\lambda}^2}, & 2 < r < n-m, \\ & r\ even. \end{cases}$$

$$\nu_{n-m} \stackrel{\text{def}}{=} \begin{cases} -\dfrac{n\nu_{r-1}}{\bar{\lambda}} + \sum\limits_{j=n-m}^{n} |a_{i,j}| + \dfrac{\bar{\lambda}|x_{n-m+i}(0)|}{\Delta_F^{(max)}}, \\ if \quad n-m\, odd; \\ \dfrac{n^2\nu_{r-2}}{\bar{\lambda}^2} + \sum\limits_{j=n-m}^{n} |a_{i,j}| + \dfrac{\bar{\lambda}|x_{n-m+i}(0)|}{\Delta_F^{(max)}}, \\ if \quad n-m\, even. \end{cases} \tag{4.64}$$

and choose $d_{i,r}$ such that $(k = r - \ell - 1)$:

$$\bar{\lambda}|d_{i,r}| < \frac{1}{1 + 2\xi_r(\mathbf{x}(0))} \cdot \Big\{ |\varepsilon_i| \left[M_1 \mu_r - \nu_r \right] - \bar{\lambda}|c_{i,r}|\xi_r(\mathbf{x}(0))$$
$$+ \sum_{\ell=1}^{r-1} |c_{i,\ell}| (-n)^{k+1} g_\ell(\mathbf{x}(0)) \left(\frac{1}{\bar{\lambda}^k} \right)^k \Big\} \tag{4.65}$$

Again, consistency requires that:

$$|\varepsilon_i| > \frac{-\sum\limits_{\ell=1}^{r-1} |c_{i,\ell}| (-n)^{k+1} g_\ell(\mathbf{x}(0)) \left(\frac{1}{\bar{\lambda}} \right)^k + \bar{\lambda}|c_{i,r}|\xi_r(\mathbf{x}(0))}{(M_1 \mu_r - \nu_r)} \stackrel{\text{def}}{=} Q_{i,r}$$

and:

$$M_1 \mu_r - \nu_r > 0 \tag{4.66}$$

This last inequality can be shown to imply (see Lemma A.1 in the Appendix) the following conditions:

$$0 < \bar{\lambda} < \bar{\lambda}_r, \quad for\ r\ even \quad \bar{\lambda} > \bar{\lambda}_r > 0, \quad for\ r\ odd \tag{4.67}$$

In Lemma A.1, reported in the Appendix, it will be shown that a suitable $\bar{\lambda}$ satisfying (4.66) can always be found. Taking into account condition (4.65), the general inequality (4.54) reads:

$$
\begin{aligned}
& -\sum_{\ell=1}^{r} |c_{i,\ell}| \,(-n)^{k+2} g_\ell(\mathbf{x}(0)) \left(\frac{1}{\bar{\lambda}^k}\right)^{k+1} + \frac{|\varepsilon_i|}{\bar{\lambda}} (M_1 \mu_r - v_r) \\
& + \sum_{j=r+1}^{n} \left[\bar{\lambda}\left(|c_{i,j}| + 2|d_{i,j}|\right) \xi_r(\mathbf{x}(0)) + |\varepsilon_i||a_{i,j}|\right] \\
& + \sum_{j=r+1}^{n-m} \left[n\left(|c_{i,j}| + |d_{i,j}|\right) + \bar{\lambda}|d_{i,j}|\right] \leq M_1 |\varepsilon_i| \sum_{j=r+1}^{n-m+1} m_j^{-1} \\
& - \begin{cases} |\varepsilon_1||a_{i,r}| & if\ r\ even \\ 0 & if\ r\ odd \end{cases}
\end{aligned} \tag{4.68}
$$

Step $n-m$. To investigate the last step, consider $r = n-m$ in the previous inequality, and recall that $|d_{i,j}| = 0$ for $j > n-m$. Moreover, $|c_{i,j}| = |\varepsilon_i|$ for $j = n-m+i$ and $|c_{i,j}| = 0$ otherwise. Considering (4.65), (4.63) (4.64), what remain of (4.54) is: $n|c_{i,n-m}| + n|d_{i,n-m}| < M_1|\varepsilon_i|m_{n-m+1}^{-1}$, i.e. $|d_{i,n-m}| < M_1 \dfrac{|\varepsilon_i|}{n \cdot m_{n-m+1}} - |c_{i,n-m}|$. Note that this latter is the second condition imposed on $|d_{i,n-m}|$, and needs to be satisfied together with (4.65) for $r = n-m$. The last condition requires $|\varepsilon_i| > \dfrac{n|c_{i,n-m}|m_{n-m+1}}{M_1} \overset{\text{def}}{=} Q_{i,n-m+1}$.

Summing up, all components of the control input v_i can be computed choosing parameter $\bar{\lambda}$ satisfying conditions (4.66), (4.67), parameters $|d_{i,j}|$, $j = 1,\ldots,n-m$, fulfilling conditions (4.56), (4.65), and parameter ε_i such that:

$$
|\varepsilon_i| > \max\{Q_{i,1}, Q_{i,2}, \ldots, Q_{i,n-m+1}\} \overset{\text{def}}{=} Q_i \quad i = 1,\ldots,m. \tag{4.69}
$$

□

4.2.3 Some Remarks

Theorem 4.6 has a *semiglobal* validity if the plant is assumed having all the eigenvalues with strictly negative real part, otherwise it holds for a compact set of initial states, known as null controllable region. Explicit characterization and description of such regions are available in the literature [15] for completely known plants, and one can expect that the recoverable region in the presence of uncertainties is a subset of the previous one. In the case of the proposed control technique, the shape of the null controllable region is also tied to the choice of the parameters ε_i, as discussed in the following. Some heuristic comments, obtained by simulation, have been added in the Results section addressing the recoverable region which can be dealt with the proposed time-varying controller.

Analyzing the dependence of the bound $\Delta_F^{(max)}$ on the tuning parameters of the control algorithm, it can be easily proved that the bound $\Delta_F^{(max)}$ depends both on the parameter $\bar{\lambda}$ and on the parameters ε_i. Such dependence can be qualitatively studied considering the dynamics of the reduced order system (A.2) (see Appendix), and evaluating the behavior of the solutions of the differential equation. As far as the dependence on $\bar{\lambda}$ is concerned, integrating the differential equation and taking the worst case, what turns out is that $\Delta_F^{(max)}(\bar{\lambda})$ depends on $\bar{\lambda}$ as $\exp(\frac{1-\exp(-\bar{\lambda}t)}{\bar{\lambda}})$, this showing that it tends to ∞ for $\bar{\lambda}$ tending to zero for any finite t, as one could expect. Following an analogous approach, it can be qualitatively found that the dependence of $\Delta_F^{(max)}$ on ε is of the form $\frac{1}{\varepsilon}$. The described dependence $\Delta_F^{(max)}(\frac{1}{\varepsilon})$ implies that $\forall\varepsilon > \bar{\varepsilon}$ it holds $\Delta_F^{(max)}(\frac{1}{\varepsilon}) < \Delta_F^{(max)}(\frac{1}{\bar{\varepsilon}})$. It follows that, fixed an a priori value $\bar{\varepsilon}$ and accordingly $\Delta_F^{(max)}(\frac{1}{\bar{\varepsilon}})$, any ε produced by the algorithm is admissible due to the inequality (4.69), since the corresponding $\Delta_F^{(max)}(\frac{1}{\varepsilon})$ is smaller than $\Delta_F^{(max)}(\frac{1}{\bar{\varepsilon}})$.

4.2.4 A Procedure for Determining the Coefficients of the Sliding Surface

The tuning of coefficients $d_{i,j}$, $j = 1,\ldots,n-m$, $i = 1,\ldots,m$ in (4.47) should be done following the lines of the proof of Theorem 4.6. Operatively, one need to compute first the upper limit $\bar{\lambda}_M$ for $\bar{\lambda}$ using plant coefficients. Next, once the parameter γ has been fixed, parameters m_j, $j = 1,\ldots,n-m+1$ can be determined, and the lower bound $\bar{\lambda}_m$ can be computed. Finally, choosing an admissible $\bar{\lambda}$ one can determine the parameters ε_i, $i = 1,\ldots,m$ and, accordingly, proper bounds for coefficients of the $\mathbf{D}$ matrix can be obtained. Note that the parameter γ (see Lemma A.1) fixed a priori could be inconsistent with the steps to follow, and could be necessary to increase it and repeat the procedure. This strategy can be expressed as a step-by-step procedure as follows:

1. Fix $i = 1$, and compute $\bar{\lambda}_M$ in (A.4) using $\bar{\lambda}_j$ with j even (only plant parameters are needed).
2. Set an arbitrary $\gamma > 1$, and fix the largest odd m_k, $k < m-n+1$, and all even m_j's according to $m_k = m_j = n$.
3. Starting from $j = k$, find the bound $m_{k-2,m}$ on m_{k-2} ensuring $\bar{\lambda}_k < \bar{\lambda}_M$.
4. Set $m_{k-2} > \max\{n, m_{k-2,m}\}$
5. Proceeding in reverse order from the largest to the smallest odd m_i, repeat the previous two steps.
6. Once all the m_i's have been computed, determine $\bar{\lambda}_m$.
7. Check if (A.7) is fulfilled. If this is not the case, increase γ and go back to step 3. The procedure is guaranteed to converge, since the increase of γ produces a corresponding linear increase of $m_{1,m}$, $m_{3,m}$, ..., while any $\bar{\lambda}_j$, j odd $\neq k$ goes to ∞ as $(\sqrt{\gamma})$.
8. Once a suitable γ has been found, choose $\bar{\lambda}$ within the interval $[\bar{\lambda}_m\ \bar{\lambda}_M]$.
9. Choose ε_i according to (4.69)
10. Finally, select $d_{i,j}$ according to the inequalities (4.56), (4.57), etc. and iterate for $i = 2,\ldots,m$.

4.2.5 Practical Issues

To ease the application of the suggested method, some hints about the practical application of the procedure will be given in the following:

- A first important choice consists in selecting ε_i large enough, in view of the presence of a lower bound only. With this choice, the feasibility range of all $d_{i,j}$ can be arbitrarily widened (in other words, $d_{i,j}$ can be almost arbitrarily fixed as initial guesses).
- In fact, since $(\mathbf{CB})^{-1}$ tends to the zero matrix when ε_i tends to infinity, the control activity can be partially reduced by increasing ε_i.
- A further important issue is to choose initial guesses of the odd m_i's (excluding the largest) not too larger than n. This makes easier the fulfillment of (A.7) and the finding of a bound for γ close to 1.
- Since the parameter $\bar{\lambda}_M$ can be computed exactly using the plant parameters, one should choose initial guesses of the odd m_i's such that the quantity $\frac{n^2}{m_i}$ is significantly lower than $\bar{\lambda}_M$, in order to enlarge the feasibility interval for $\bar{\lambda}$.

4.2.6 Simulation Results

The proposed control approach, based on sliding surface (4.47) and control law (4.48), has been applied by simulation to an unstable 2-inputs plant the form (4.40) with: $n = 4$, $m = 2$, $a_{1,1} = 1; a_{1,2} = -0.2; a_{1,3} = 0.3; a_{1,4} = 0$; $a_{2,1} = a_{2,4} = 1; a_{2,2} = a_{2,3} = -1$. The control input $\mathbf{u}$ feeding the plant is the output of a saturation device, with threshold $M = 3$. A disturbance term of the form $d(t) = A\sin(\omega t)$ has been supposed to perturb the system, with $|A| \le 0.1$ and $\omega = 1$. Following the procedure described in Section 4.2.4, having set $\rho_1 = \rho_2 = 2$, we found the following parameters: $\gamma = 1.1$, $m_1 = m_2 = m_3 = 4$. The matrix $\mathbf{C}$ has been selected as $\mathbf{C} = \begin{bmatrix} 3000 & 20000 & 155340 & 0 \\ 1 & 1 & 0 & 600 \end{bmatrix}$ while the entries of the matrix $\mathbf{D}$ have been designed according to Theorem 4.6. Found maximal values are $\mathbf{D} = \begin{bmatrix} 0.0795 & 0.5084 & 0 & 0 \\ 5.1647 & 0.8541 & 0 & 0 \end{bmatrix} 10^4$. The value of $\mathbf{D}$ used in simulations were obtained scaling each entry by a factor $-1/80$. In order to avoid chattering, a boundary layer has been introduced [58]. Simulations have been performed with $\bar{\lambda} = 3$ and initial condition $\mathbf{x}(0) = [0.5\ 0.5\ 0\ 0]^T$. Note that no constraints are required between the bounding function $\Delta_F^{(max)}$ and the threshold M, as it happened in [40]. We could set $\Delta_F^{(max)} = 2.8$ as needed, regardless the value of M. Results have been reported in Fig. 4.10, showing the first two state variables, and in Fig. 4.11, which displays the available control inputs $\mathbf{v}$.

Performances of the proposed controller have been compared by simulation with a standard linear state feedback controller (assigning eigenvalues -0.1, -0.2, -0.3, -0.4). It was found that, in the presence of the disturbance, the set of initial

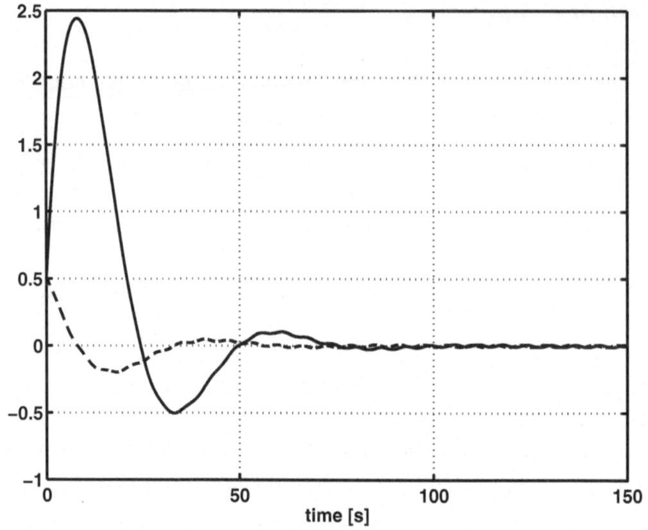

Fig. 4.10. State variable x_1 (solid line) and x_2 (dashed line)

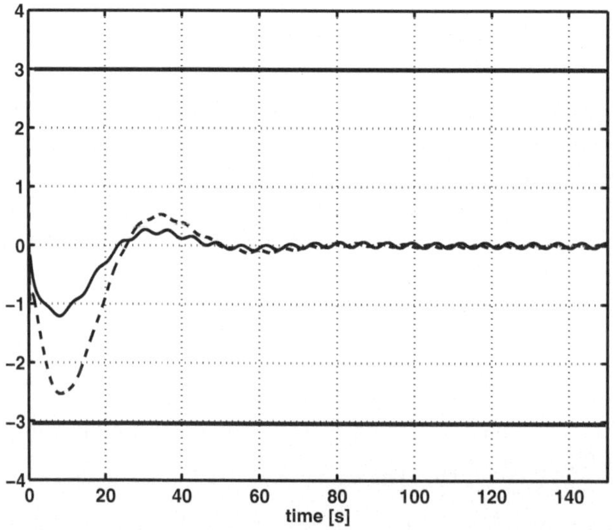

Fig. 4.11. Unavailable Control inputs u_1 (solid line) and u_2 (dashed line) with the threshold $\pm M$

conditions from which the plant could be steered to the origin was larger for the time-varying controller. Just as an example, the initial condition $\mathbf{x}(0) = [1\ 0.65\ 0\ 0]^T$ could be effectively managed by the proposed control technique (maintaining the settings described) but caused instability using the standard controller.

Chapter 5
Control Design Issues: Discrete-Time Plants

This chapter is devoted to the extension of the results reported in the previous Chapter 4 to the discrete-time framework. Needless to say, completely different proofs are required with respect to the continuous time case. As in the previous Chapter, a discrete-time linear n-dimensional plant without uncertainties is considered first, and a design technique is proposed for finding a linear state feedback controller with the property of having non-increasing norm along the closed-loop system trajectories.

In the presence of disturbances affecting the plant, we are again concerned on achieving ultimate boundedness of trajectories and ensuring disturbance rejection. In view of the widely recognized features of robustness, transient shaping and ease of application offered by the discrete-time counterpart of sliding modes [59] [56], often referred to as quasi-sliding modes, one should expect their use to be widespread also for ensuring transient performance requirements and robust stabilization in the case of discrete-time plants with saturating actuators, too. Nevertheless, very few results are available.

In the vast literature addressing the stabilization problem for discrete-time linear systems subject to actuator saturation, two lines of research have been mostly pursued, as discussed in the present book. The first line focusses on the the estimation, less conservative as possible, of the null controllable region, i.e. the set of state which can be driven towards the origin of the state space using saturating actuators [60], [61], [62], [63], [64]. This research thrust has been widely addressed in Part I. The other line of research focuses on the semi-global stabilization on the null controllable region using saturating actuators. In this latter framework, the problem has been completely studied for ANCBC plants, for which the null controllable region is the whole state space [10], [11], [65]. In particular, some results are available for general discrete-time systems about feedback laws achieving semi-global stabilization on the null controllable region [66], [67]. However, [68] showed that in general linear feedback can not achieve global stabilization for discrete-time unstable systems.

The problem of disturbance rejection for linear systems subject to actuator saturation has been investigated also in the discrete time framework, mostly considering

M.L. Corradini et al.: Control Systems with Saturating Inputs, LNCIS 424, pp. 95–124.
springerlink.com

disturbances that are bounded in magnitude. Also, the problem of the robust semi-global [15] and global [23] stabilization has been solved for planar continuous-time systems with two unstable open-loop poles, in the case when the constraints are posed on the inputs. A more general setting including input and state constraints has been studied in [69], where solvability conditions have been provided for semi-global stabilization in the admissible set of systems subject to right-invertible and non-minimum phase constraints. Furthermore, it has been proved [70] that for a discrete-time linear system with saturating actuators, neither L_q semiglobal L_p-stabilization nor global or semiglobal external stabilization is possible whenever there is one controllable open-loop pole located strictly outside the unit circle. Finally, it has been shown [71] that for both continuous-time and discrete-time critically unstable linear systems with saturating actuators, one cannot simultaneously achieve the global internal stabilization (in the absence of disturbance) and the global finite-gain L_p performance in the presence of disturbance whenever the external input (disturbance) is not input-additive. However, one can achieve the global internal stabilization and global L_p stabilization (without finite-gain) for any $p \in [1,\infty)$. Estimation methods of the domain of attraction in the uncertain case using a saturation-dependent Lyapunov function has been proposed in [64] [72]. Also, the enlargement of the domain of attraction by antiwindup compensation has been proposed [73].

In addition to simple stabilization, the issue of guaranteeing some performance requirements has been also addressed mostly using LMI's for completely known discrete-time plants [74] [60], [67] and for uncertain plants [75], but with reference to particular classes of disturbances. These approaches basically convert constraints to the form of LMI's, so that the eventual solution to the control problem is conditioned to the feasibility of a convex problem.

The extension of the design technique presented in Chapter 4 to the discrete-time framework presents some relevant features: *i*) no restrictions are needed in the plant structure; *ii*) bounded matched disturbances are considered; *iii*) robust practical *semi-global stabilization on the null controllability region* (as defined in [10] [23]) can be achieved by means of a time-varying state feedback controller, derived imposing the achievement of a quasi-sliding motion onto a suitable time-varying sliding surface; *iv*) performing transient shaping is not subject to any condition and can be achieved simply by manipulating the dynamics imposed onto the quasi-sliding surface. It will be proved that a constructive systematic procedure exists for designing the surface as to guarantee the ultimately boundedness of plant trajectories in the presence of bounded matched uncertainties. Finally, results from experiments in a twin rotor system will be reported and discussed.

5.1 Invariant Strips and Linear Feedback Laws

In this section we explain how to extend the previous results to discrete linear systems.

We consider a n-dimensional controllable SISO linear discrete system having, without loss of generality, the following structure

$$\begin{cases} \mathbf{x}(k+1) = \mathbf{A}\mathbf{x}(k) + bu(k) \\ \mathbf{x}(0) = \mathbf{x}_0 \end{cases} \tag{5.1}$$

where

$$\mathbf{A} = \begin{pmatrix} 0 & 1 & \cdots & 0 \\ \vdots & \vdots & \vdots & \vdots \\ 0 & \cdots & 0 & 1 \\ a_1 & a_2 & \cdots & a_n \end{pmatrix}, \qquad b = \begin{pmatrix} 0 \\ \vdots \\ 0 \\ 1 \end{pmatrix},$$

$\{\mathbf{x}(k)\}_{k\in\mathbb{N}} \subset \mathbb{R}^n$ is the state vector and $\{u(k)\}_{k\in\mathbb{N}} \subset \mathbb{R}$ is the input variable, which is subject to the saturation constraint $u(k) = \text{sat}_M(v(k)),\ M > 0$.

Denote by $\{\lambda_j\}$ the eigenvalues of the matrix $\mathbf{A}$. It will be supposed that

$$|\lambda_j| < 1 \quad \forall\, 1 \le j < n,\ \lambda_n \in \mathbb{R} \text{ with } |\lambda_n| \ge 1. \tag{5.2}$$

In the next result it will be discussed how to design a linear feedback such that the set of values satisfying the saturation constraint is invariant for the closed-loop system evolution.

Lemma 5.1. *Given the system (5.1) controlled by the linear feedback* $u = \langle \mathbf{l}, \mathbf{x} \rangle$ *with* $\mathbf{l} \in \mathbb{R}^n$*, the following propositions are equivalent:*

1. $|\langle \mathbf{l}, \mathbf{x}(k) \rangle| \le M \Rightarrow |\langle \mathbf{l}, \mathbf{x}(k+1) \rangle| \le M,\ \forall k \in \mathbb{N},\ \forall M > 0;$
2. $\mathbf{l}$ *is an eigenvector of the matrix* $\mathbf{A}^T$ *and the corresponding eigenvalue* λ *satisfies* $|\lambda + l_n| \le 1$*, where* l_n *is the* n^{th} *component of* $\mathbf{l}$.

Proof. Given $M > 0$, we consider the set $\Omega_M = \{x \in \mathbb{R}^n : |\langle \mathbf{l}, \mathbf{x} \rangle| \le M\}$; our aim is to construct $\mathbf{l}$ such that $\mathbf{x}(k) \in \Omega_M$ implies $\mathbf{x}(k+1) \in \Omega_M$, that is

$$|\langle \mathbf{l}, \mathbf{A}\mathbf{x}(k) + b\mathbf{l}^T\mathbf{x}(k) \rangle| \le M. \tag{5.3}$$

Rewriting the previous condition as

$$|\langle \mathbf{A}^T\mathbf{l}, \mathbf{x}(k) \rangle + l_n \langle \mathbf{l}, \mathbf{x}(k) \rangle| \le M.$$

we see that the invariance of Ω_M can be achieved if and only if $\mathbf{l}$ is an eigenvector of the matrix $\mathbf{A}^T$ with a corresponding real eigenvalue μ such that

$$|\mu \langle \mathbf{l}, \mathbf{x}(k) \rangle + l_n \langle \mathbf{l}, \mathbf{x}(k) \rangle| \le M.$$

We can conclude that condition (5.3) holds if and only if we have

$$|\mu + l_n| \le 1. \tag{5.4}$$

$\diamond$

We can state the following theorem, which can be regarded as a discrete version of Theorem 4.2.

Theorem 5.1. *Given the system (5.1) satisfying condition (5.2), there exists a vector* $\bar{\mathbf{l}} \in \mathbb{R}^n$ *such that for any* $M > 0$

1. $\Omega_M = \{\mathbf{x} \in \mathbb{R}^n : |\langle \bar{\mathbf{l}}, \mathbf{x} \rangle| \leq M\}$ *is an invariant set for the solution* $\{\phi(k, \mathbf{x}_0, \bar{\mathbf{l}})\}_{k \in \mathbb{N}}$*;*
2. *if* $\mathbf{x}_0 \in \Omega_M$ *we have* $\lim_{k \to \infty} \|\phi(k, \mathbf{x}_0, \bar{\mathbf{l}})\| = 0$ *under the saturation constraint (2.2).*

Proof. Let us define $\bar{\mathbf{l}} = \sigma \mathbf{l}^{(n)}$, where $\mathbf{l}^{(n)}$ is an eigenvector of the matrix $\mathbf{A}^T$ associated to the eigenvalue λ_n and σ is a real parameter to be determined. Recalling result of Lemma 4.2 and since assumption (5.2) holds, we can obtain asymptotical stabilization of the system choosing σ such that

$$|\lambda_n + \sigma l_n^{(n)}| < 1. \tag{5.5}$$

Moreover, by Lemma 5.1 condition (5.5) is sufficient also for ensuring invariance of the sub-level sets of the control norm $|u|$. ◊

5.2 Nonlinear Robust Controller Design via Quasi-sliding Modes

5.2.1 Problem Statement

Consider the following time invariant, uncertain, discrete time, controllable multi-input plant given by:

$$\mathbf{x}(k+1) = \mathbf{A}\mathbf{x}(k) + \mathbf{B}\left[\mathbf{sat}(\mathbf{u}(k)) + \delta(k)\right] \tag{5.6}$$

where: $\mathbf{x} = [x_1(k) \cdots x_n(k)]^T \in \mathbb{R}^n$ is the state vector (assumed available for measurement), $\mathbf{u}(k) \in \mathbb{R}^m$ is the control input, $\mathbf{A} \in \mathbb{R}^{n \times n}$, $\mathbf{B} \in \mathbb{R}^{n \times m}$ are the state matrix and the input matrix respectively. The uncertain term $\delta(k) \in \mathbb{R}^m$ represent matched disturbances affecting the system. Given the standard saturation function $\mathbf{sat}(\cdot)$ with threshold $M > 0$ known, the i-th component of the vectorial function $\mathbf{sat}(\cdot) : \mathbb{R}^m \to \mathbb{R}^m$ is given by $sat(u_i)$. Without loss of generality, the plant is given in a form obtained rearranging rows and columns of the controllable canonical form, which depends on the controllability indices of the system n_j, $j = 1 \ldots m$ [54] [55]: $\mathbf{A} = \begin{bmatrix} \mathbf{A}_1 \\ \mathbf{A}_2 \end{bmatrix}$, $\mathbf{B} = \begin{bmatrix} 0_{(n-m) \times m} & I \end{bmatrix}^T$, with $\mathbf{A}_1 \in \mathbb{R}^{(n-m) \times n}$, $\mathbf{A}_2 \in \mathbb{R}^{m \times n}$, where the block matrix $\mathbf{A}_1$ has elements $\{\xi_{i,j}\}$, $i = 1, \ldots, n-m$, $j = 1, \ldots, n$ which are either 1 or 0, and

$$\mathbf{A}_2 = \begin{bmatrix} a_{1,1} & a_{1,2} & \cdots & a_{1,n} \\ & & \cdots & \\ a_{m,1} & a_{m,2} & \cdots & a_{m,n} \end{bmatrix}$$

Assumption 5.1. *The uncertain term $\delta(k)$ is bounded by a known constant, i.e.* $||\delta(k)|| \leq \rho$.

Definition 5. Denote solutions of a general system $\mathbf{x}(k+1) = f(\mathbf{x}(k),k)$ as $\phi(k,k_0,\mathbf{x}(0))$ with initial condition $\mathbf{x}(0)$. Following [76], such solutions are defined *uniformly ultimately bounded* (with bound B) if there exists a $B > 0$ and if corresponding to any $\alpha > 0$ and for every $k_0 \in \mathbb{N}$, there exists a $T = T(\alpha) > 0$ (independent of k_0) such that $||\mathbf{x}(0)|| < \alpha$ implies that $||\phi(k,k_0,\mathbf{x}(0))|| < B$ for all $k \geq k_0 + T(\alpha)$.

Define the matrices $\mathbf{C} = \begin{bmatrix} \Gamma \ \mathbf{E} \end{bmatrix} \in \mathbb{R}^{m\times n}$ and $\mathbf{D} = \begin{bmatrix} \Delta \ 0 \end{bmatrix} \in \mathbb{R}^{m\times n}$ as

$$\mathbf{D} = \begin{bmatrix} d_{11} & d_{12} & \dots & d_{1,n-m} & 0 & 0 \\ & & \dots & & \dots & \\ d_{m,1} & d_{m,2} & \dots & d_{m,n-m} & 0 & 0 \end{bmatrix} \quad \mathbf{C} = \begin{bmatrix} c_{11} & c_{12} & \dots & c_{1,n-m} & \varepsilon_1 & 0 & \dots 0 \\ & & \dots & & \dots & & \\ c_{m,1} & c_{m,2} & \dots & c_{m,n-m} & 0 & \dots 0 & \varepsilon_m \end{bmatrix} \tag{5.7}$$

and $\bar{\mathbf{C}}(k) \stackrel{\text{def}}{=} (\mathbf{C} + \mathbf{D}\bar{\lambda}^k) = [\bar{c}_{i,j}]$. Consider the following time-varying sliding surface, with $|\bar{\lambda}| < 1$:

$$\mathbf{s}(\mathbf{x}(k),\mathbf{x}(0),k) = \bar{\mathbf{C}}(k)\left[\mathbf{x}(k) - \mathbf{x}(0)\bar{\lambda}^k\right] = 0 \tag{5.8}$$

It will be proved in the following that, for any choice of $d_{i,j} \lesseqgtr 0$, $i = 1,m$, $j = 1,\dots,n-m$, a quasi sliding motion onto the surface $\mathbf{s}(\mathbf{x}(k),\mathbf{x}(0),k) = 0$ implies ultimate boundedness of state trajectories.

What motivates the introduction of the vanishing term $\mathbf{D}\bar{\lambda}^k\mathbf{x}$ with respect to standard surfaces is the need of modulating the control input in order to cope with the saturation limitation. The matrix $\mathbf{C}$, as usual, is chosen as to assign the eigenvalues of the reduced order system (see proof of Theorem 5.3 in Section 5.2.4). Recall that the robust stabilization of the reduced order system can be achieved in the presence of matched disturbances [50].

5.2.2 The Control Law

For the surface (5.8), the control input ensuring the achievement of a quasi sliding motion is [56]:

$$\begin{aligned} \mathbf{u}(k) &= -[\bar{\mathbf{C}}(k+1)\mathbf{B}]^{-1}\left[\mathbf{C}\mathbf{A}\mathbf{x}(k) + \mathbf{D}\bar{\lambda}^{k+1}\mathbf{A}\mathbf{x}(k) - \bar{\mathbf{C}}(k+1)\mathbf{x}(0)\bar{\lambda}^{k+1} + \mathbf{u}^{(n)}(k)\right] \\ &= -\mathbf{E}^{-1}\left\{\left[\left(\Gamma + \Delta\bar{\lambda}^{k+1}\right)\mathbf{A}_1 + \mathbf{E}\mathbf{A}_2\right]\mathbf{x}(k) - \bar{\mathbf{C}}(k+1)\mathbf{x}(0)\bar{\lambda}^{k+1} + \mathbf{u}^{(n)}(k)\right\} \end{aligned} \tag{5.9}$$

with

$$u_i^{(n)}(k) = \begin{cases} -\theta_i(|s_i(k)| - \rho) & if |s_i(k)| \geq \rho \\ 0 & if |s_i(k)| < \rho \end{cases} \qquad |\theta_i| \leq 1 \quad i = 1,\dots,m \tag{5.10}$$

where, with some abuse of notation, the i-th component of the variable $\mathbf{s}(\mathbf{x}(k),\mathbf{x}(0),k)$ has been denoted by $s_i(k)$. Control law (5.9)-(5.10) is obtained solving $\mathbf{s}(k+1)=\mathbf{0}$ in nominal conditions, and then choosing $u_i^{(n)}(k)$ in order to impose $||\mathbf{s}(k+1)|| < ||\mathbf{s}(k)||$ [56].

The following Proposition is straightforward, in view of the absence of the reaching phase.

Proposition 5.1. *It is given the uncertain system (5.6) driven by the feedback controller (5.9) under Assumption 5.1 and the assumption that $M > \rho$. For any bounded set $\mathscr{S}$ of initial conditions $\mathscr{S} = \{\mathbf{x}(0) : ||\mathbf{x}(0)|| \leq \zeta;\ \zeta > 0\}$ belonging to the null controllability region, there exists a constant $\Delta_F^{(max)} \in \mathbb{R}^+$, depending on the chosen set $\mathscr{S}$ and on the parameter $0 < |\lambda| < 1$, such that $\sup_{\mathbf{x}(0)\in\mathscr{S},0<|\bar{\lambda}|<1} ||\mathbf{x}(k)|| \leq \Delta_F^{(max)}$, $\forall k \geq 0$.*

The fact that the initial conditions are assumed to belong to the null controllability region in Proposition 5.1 ensures that the state can be driven to the origin using bounded control inputs, this ensuring boundedness of the state of the saturated plant.

Remark 5.1. *The bound $\Delta_F^{(max)}$, which always exists for initial conditions belonging to the null controllability region, can be estimated exploiting available boundedness results for closed loop trajectories (see Section IV, eq. (5.40)). It should be noticed that it is not necessary to achieve an accurate estimate of $\Delta_F^{(max)}$, since even a rough overestimated upper bound of trajectories can be enough for the determination of the coefficients of the time varying sliding surface.*

The constraint induced by saturation requires:

$$|u_i(k)| \leq M \quad \forall k \geq 0, \quad i = 1,\ldots,m \tag{5.11}$$

Taking the worst case and considering (5.9) it follows that the condition (5.11) can be rewritten as:

$$|\varepsilon_i u_i(k)| = \Bigg|\bar{\lambda}^{k+1}\sum_{p=1}^{n-m}\left(c_{i,j}+d_{i,j}\bar{\lambda}^{k+1}\right)x_j(0)+\varepsilon_i x_{n-m+i}(0)\bar{\lambda}^{k+1}-\varepsilon_i\sum_{j=1}^{n}a_{i,j}x_j(k)+ \tag{5.12}$$

$$-\sum_{p=1}^{n-m}\bar{c}_{i,j}(k)\sum_{\ell=1}^{n}\xi_{j,\ell}x_\ell(k)-\theta_i|s_i(k)|\Bigg| \leq \bar{M}|\varepsilon_i| \tag{5.13}$$

where $\bar{c}_{i,j}(k) = c_{i,j}+d_{i,j}\bar{\lambda}^k$, $j = 1,\ldots,n-m$, $\bar{M} \overset{\text{def}}{=} M-|\theta_i|\rho$. Considering the expression of $|s_i(k)|$, and taking again the worst case, one has:

$$\varepsilon_i\left[\sum_{j=1}^{n}|a_{i,j}|+\bar{\lambda}+2\theta_i\right]+\gamma_i\sum_{p=1}^{n-m}|c_{i,j}|+2\bar{\lambda}\sum_{p=1}^{n-m}|d_{i,j}|+2\sum_{p=1}^{n-m}|d_{i,j}| \leq M_1\varepsilon_i \tag{5.14}$$

with $M_1 = \dfrac{\bar{M}}{\Delta_F^{(max)}}$, $\sigma_1 = \bar{\lambda} + 1$, $\gamma_i = \sigma_1 + 2\theta_i$. The following Theorem provides a robust stabilizing controller designed as to fulfill the constraint $|u_i(k)| \le M$ associated to saturation. It will be given under the standard assumption that the plant is ANCBC (i.e., the eigenvalues of $\mathbf{A}$ are on or inside the unit circle), which is necessary for semiglobal stabilization. Without such an assumption on the open-loop eigenvalues, only local results are possible.

Theorem 5.2. *It is given the uncertain system (5.6), assumed ANCBC, preceded by the saturating device (4.41), under the assumption that* $\bar{M} > 0$ *and Assumption 5.1. For any given* $\mathbf{x}(0)$ *belonging to the null controllability region, proper coefficients* $d_{i,j}$, $i = 1,\ldots,m$, $j = 1,\ldots,n-m$ *and a suitable* $0 < |\bar{\lambda}| < 1$ *can always be found such that the feedback controller (5.9) guarantees that plant trajectories are semiglobally uniformly ultimately bounded.*

Proof. The proof is constructive, and consists in $n-m$ steps, providing $n-m$ chained inequalities, for each i-th component of $\mathbf{u}(k)$. As it will be clear soon, the construction is systematic, since for the i-th component the same development is applied to all the terms with subscript j at the step j-th. Moreover, odd and even steps have to be treated differently. Define $m_j > 0$, $j = 1,\ldots,m$ such that

$$\sum_{p=1}^{n-m} m_j^{-1} \le 1. \tag{5.15}$$

In the following, the imposition of condition (5.14) will be performed taking suitable $n-m$ "portions" of (5.14) itself, and designing control coefficients $d_{i,j}$, m_j, ε_i, $\bar{\lambda}$ involved in each derived inequality in order to ensure the simultaneous fulfillment of all of them.

- Consider first $j = 1$. A stronger inequality than (5.14) is the following:

$$\varepsilon_i \left[|a_{i,1}| + \bar{\lambda} + 2\theta_i\right] + \gamma_i|c_{i,1}| + 2\bar{\lambda}|d_{i,1}| + 2|d_{i,1}| + G_{i2} \le M_1\varepsilon_i \left(\frac{1}{m_1} + \sum_{p=2}^{n-m} \frac{1}{m_p}\right) \tag{5.16}$$

with $G_{i,\ell} = \varepsilon_i \sum_{p=\ell}^{n} |a_{i,j}| + \sum_{p=\ell}^{n-m} (\gamma_i|c_{i,j}| + 2\sigma_1|d_{i,j}|)$, $\ell = 2,\ldots,n$. Choose $d_{i,1}$ such that:

$$2|d_{i,1}| < \varepsilon_i\left(M_1\mu_1 - \nu_1\right) - \gamma_i|c_{i,1}| + K_1 \tag{5.17}$$

where $\mu_1 = \dfrac{1}{m_1}$, $\nu_1 = |a_{i,1}| + \bar{\lambda} + 2\theta_i$, and a same constant K_1, to be determined, has been added and subtracted in (5.16). For the r.h.s. term of condition (5.17) to be positive, one can require: $m_1 > \dfrac{M_1}{\nu_1}$, $K_1 > \gamma_i|c_{i,1}| \stackrel{\text{def}}{=} K_1^*$ and

$$|\varepsilon_i| < \frac{K_1 - \gamma_i|c_{i,1}|}{\nu_1 - M_1\mu_1} \stackrel{\text{def}}{=} Q_{i,1}. \tag{5.18}$$

- Consider $j = 2$. Replacing (5.17) in (5.16) one gets:

$$K_1\sigma_1+\gamma_i\left(|c_{i,2}|-\bar{\lambda}|c_{i,1}|\right)+2\bar{\lambda}|d_{i,2}|+2|d_{i,2}|+G_{i3}\leq \varepsilon_i\left(M_1\mu_2-\nu_2\right)+M_1\varepsilon_i\sum_{p=3}^{n-m}\frac{1}{m_p} \tag{5.19}$$

being:

$$\mu_j \overset{\text{def}}{=} \frac{1}{m_j}-\bar{\lambda}\mu_{j-1};\quad \nu_j \overset{\text{def}}{=} |a_{i,j}|-\bar{\lambda}\nu_{j-1} \tag{5.20}$$

Choose $d_{i,2}$ such that:

$$2|d_{i,2}| < \varepsilon_i\left(M_1\mu_2-\nu_2\right)+\gamma_i\alpha_{i,2}+K_2-\sigma_1 K_1 \tag{5.21}$$

with $\alpha_{i,\ell} = \sum_{j=1}^{\ell}(-1)^{j+\ell+(\ell \, mod \, 2)}\cdot\bar{\lambda}^{j-1}|c_{i,(\ell+1-j)}|$. Condition (5.21) requires:

$$\varepsilon_i < \frac{K_2-\sigma_1 K_1+\gamma_i\alpha_{i,2}}{\nu_2-M_1\mu_2} \overset{\text{def}}{=} Q_{i,2} \tag{5.22}$$

provided that K_2 is large enough: $K_2 > \sigma_1 K_1 - \gamma_i\alpha_{i,2} \overset{\text{def}}{=} K_2^*$, and provided that $M_1\mu_2-\nu_2 < 0$. To this purpose, one can impose $\mu_2 < 0$ and $\nu_2 > 0$, i.e.:

$$\bar{\lambda} < \bar{\lambda}_2 = \frac{|a_{i,2}|}{|a_{i,1}|+2\theta_i+1};\quad m_2 > \frac{m_1}{\bar{\lambda}}. \tag{5.23}$$

- Consider $j = 3$. Replacing (5.21) in (5.19), one gets:

$$\beta_2\sigma_1-\gamma_i\alpha_{i,3}+2\bar{\lambda}|d_{i,3}|+2|d_{i,3}|+G_{i,4}\leq \varepsilon_i\left(M_1\mu_3-\nu_3\right)+M_1\varepsilon_i\sum_{p=4}^{n-m}\frac{1}{m_p} \tag{5.24}$$

with $\beta_\ell = \sum_{j=1}^{\ell}(-1)^{1-j+\ell+(\ell \, mod \, 2)}\cdot\bar{\lambda}^{j-1}K_{(\ell+1-j)}$. Choose d_3 such that:

$$2|d_{i,3}| < \varepsilon_i\left(M_1\mu_3-\nu_3\right)+\gamma_i\alpha_{i,3}+K_3-\sigma_1\beta_2. \tag{5.25}$$

Differently from the previous case, where $\mu_2 < 0$ and $\nu_2 > 0$, the condition $M_1\mu_3-\nu_3 < 0$ has now to be imposed explicitly, i.e. $\bar{\lambda}(\nu_2-M_1\mu_2) < |a_{i,3}|-\frac{M_1}{m_3}$ where, setting $m_3 > \dfrac{M_1}{|a_3|}$ and recalling the step 2, both members are positive. Substituting the expressions (5.20) with $j = 2$, one gets

$$\bar{\lambda}\left(\bar{\lambda}\left(\frac{M_1}{m_1}-|a_{i,1}|-\bar{\lambda}-2\theta_i\right)+|a_{i,2}|-\frac{M_1}{m_2}\right) < |a_{i,3}|-\frac{M_1}{m_3} \tag{5.26}$$

and, since $m_1 > \dfrac{M_1}{\nu_1}$, a strongest condition is $\bar{\lambda}(|a_{i,2}|-\frac{M_1}{m_2}) < |a_{i,3}|-\frac{M_1}{m_3}$ which provides, for $m_2 > \dfrac{M_1}{|a_{i,2}|}$,

$$\bar{\lambda} < \frac{|a_{i,3}| - \frac{M_1}{m_3}}{|a_{i,2}| - \frac{M_1}{m_2}} \stackrel{\text{def}}{=} \bar{\lambda}_3 > 0 \tag{5.27}$$

Condition (5.25) requires:

$$\varepsilon_i < \frac{K_3 - \sigma_1 \beta_2 + \gamma_i \alpha_{i,3}}{(\nu_3 - M_1 \mu_3)} \stackrel{\text{def}}{=} Q_3 \tag{5.28}$$

provided that K_3 is large enough: $K_3 > \sigma_1 \beta_2 - \gamma_i \alpha_{i,3} \stackrel{\text{def}}{=} K_3^*$.

- The above procedure can be generalized for any $j = r \leq n - m$. Choose $d_{i,r}$ such that:

$$2|d_{i,r}| < \varepsilon_i (M_1 \mu_r - \nu_r) + \gamma_i \alpha_{i,r} + K_r - \sigma_1 \beta_{r-1}. \tag{5.29}$$

Condition (5.29) requires:

$$\varepsilon_i < \frac{K_r - \sigma_1 \beta_{r-1} + \gamma_i \alpha_{i,r}}{(\nu_r - M_1 \mu_r)} \stackrel{\text{def}}{=} Q_{i,r} \tag{5.30}$$

and: $M_1 \mu_r - \nu_r < 0$, which implies the following conditions:

$$\begin{cases} \bar{\lambda} < \dfrac{|a_r|}{\nu_{r-1}} \stackrel{\text{def}}{=} \bar{\lambda}_r; \quad m_r > \dfrac{m_{r-1}}{\bar{\lambda}} \quad r \textit{ even} \\ \bar{\lambda} < \dfrac{|a_r| - \frac{M_1}{m_r}}{|a_{r-1}| - \frac{M_1}{m_{r-1}}} \stackrel{\text{def}}{=} \bar{\lambda}_r; \quad r \textit{ odd} \\ m_r > \dfrac{M_1}{|a_r|} \quad \forall r \end{cases} \tag{5.31}$$

$$K_r > \sigma_1 \beta_{r-1} - \gamma_i \alpha_{i,r} \stackrel{\text{def}}{=} K_r^* \quad r = 2, \ldots, n - m - 1 \tag{5.32}$$

To allow the correct ending of the procedure, the constant K_{n-m} is chosen equal to 0, and for $r = n - m$ condition (5.32) becomes: $\sigma_1 \beta_{n-m-1} < \gamma_i \alpha_{i,n-m}$. This condition can be easily imposed by using one of the redundant coefficients of the matrix $\mathbf{C}$, i.e.:

$$\gamma_i \bar{\lambda} |c_{i,n-m-1}| > \sigma_1 \beta_{n-m-1} + \gamma_i \left[|c_{i,n-m}| - \bar{\lambda}^2 \alpha_{i,n-m-2} \right] \tag{5.33}$$

- Finally, taking into account (5.29) for $r = n - m$ and (5.32), the last condition to be fulfilled in order to guarantee (5.14) is the following:

$$\bar{\lambda} \left[\varepsilon_i (M_1 \mu_{n-m} - \nu_{n-m}) + \gamma_i \alpha_{i,n-m} - \sigma_1 \beta_{n-m-1} \right] + \sum_{\ell=n-m+1}^{n} |a_{i,\ell}| < 0 \tag{5.34}$$

i.e.

$$\bar{\lambda}\sigma_1 K_{n-m-1} > \sum_{\ell=n-m+1}^{n} |a_{i,\ell}| + \bar{\lambda}\varepsilon_i(M_1\mu_{n-m} - \nu_{n-m}) + \bar{\lambda}\gamma_i\alpha_{i,n-m} + \bar{\lambda}^2\sigma_1\beta_{n-m-2} \tag{5.35}$$

which provides a further constraint on K_{n-m-1}. Note that (5.35) is the second condition imposed on K_{n-m-1}, and needs to be satisfied together with (5.29) for $r = n-m-1$. ◇

5.2.3 A Systematic Procedure

According to the proof of Theorem 5.2, the following systematic operative procedure can be given for the determination of the coefficients $d_{i,j}$, $i = 1,\ldots,m$, $j = 1,\ldots,n-m$, of the matrix $\mathbf{D}$ in (5.8).

1. Select the matrix $\mathbf{C}$ as to assign the eigenvalues of the reduced order system (see proof of Theorem 5.3 in Section 5.2.4)
2. Set $i = 1$.
3. Set $\bar{\lambda} < \min\{1, \min_i \bar{\lambda}_i, \quad i = 2,\ldots,n-m\}$;
4. Fix $m_1 = n-m$, and compute all the further m_j, $j = 2,\ldots,n-m$, according to (5.31).
5. Compute ν_j and μ_j, $j = 1,\ldots,n-m$ according to (5.20).
6. Determine numerically an (even rough) estimate of the bounding constant $\Delta_F^{(max)}$, based on the set $\mathscr{S}$ of initial conditions and the assigned eigenvalues (see Theorem 5.3 in Section 5.2.4).
7. Compute all K_j's, $j = 1,\ldots,n-m$, according to the general expression:

$$K_j > \max_j K_j^* = \max_j \{\sigma_1\beta_{j-1} - \gamma_i\alpha_{i,j}\}, \quad , \quad i = 1,\ldots,n-m-1 \tag{5.36}$$

8. Compute all $Q_{i,j}$'s, $j = 1,\ldots,n-m$, and select ε_i according to the general expression:

$$|\varepsilon_i| < \min_j Q_{i,j} = \min_j \left\{ \frac{K_j - \sigma_1\beta_{j-1} + \gamma_i\alpha_{i,j}}{(\nu_j - M_1\mu_j)} \right\} \tag{5.37}$$

9. Finally, select $d_{i,j}$, $j = 1,\ldots,n-m$, according to the general expression (5.29):

$$2|d_{i,j}| < \varepsilon_i\,(M_1\mu_j - \nu_j) + \gamma_i\alpha_{i,j} + K_j - \sigma_1\beta_{j-1}. \tag{5.38}$$

10. Repeat the whole procedure for $i = 2,\ldots,m$.

5.2.4 Stability Analysis and Transient Shaping

In the following, the ultimate boundedness of the closed loop plant trajectories is proved.

Theorem 5.3. *Consider the plant (5.6) subject to the saturation constraint (4.41). The control input (5.9) ensures that closed loop state trajectories are ultimately bounded.*

Proof. The dynamics of the plant (5.6) driven by (5.9) are:

$$\mathbf{x}(k+1) = \tilde{\mathbf{A}}\mathbf{x}(k) + \bar{\varphi}(\mathbf{x}(k), \mathbf{x}(0), k) \tag{5.39}$$

with $\tilde{\mathbf{A}} = \mathbf{A} - \mathbf{B}\left(\mathbf{E}^{-1}\Gamma\mathbf{A}_1 + \mathbf{A}_2\right)$, $\bar{\varphi}(\mathbf{x}(k), \mathbf{x}(0), k) = -\mathbf{B}\mathbf{E}^{-1}\Delta\bar{\lambda}^{k+1}\mathbf{A}_1\mathbf{x}(k) + \mathbf{E}^{-1}\bar{\mathbf{C}}(k+1)\mathbf{x}(0)\bar{\lambda}^{k+1} + \mathbf{B}\left(\mathbf{u}^{(n)}(k) + \delta(k)\right)$. Note that $|eig(\tilde{\mathbf{A}})| < 1$ due to a proper choice of Γ. Therefore $||\tilde{\mathbf{A}}^k|| \le N\lambda_M^k$, being λ_M the largest eigenvalue of $\tilde{\mathbf{A}}$, and where $N > 0$ is a suitable constant.

Since $||\bar{\varphi}(\mathbf{x}(k), \mathbf{x}(0), k)|| \le \delta_1||\mathbf{x}(k)|| + \delta_2$, with $\delta_1 = \frac{||\Delta||\bar{\lambda}}{\max\{\varepsilon_i\}}$, following the development described in [56], p. 114, one gets:

$$||\mathbf{x}(k)|| \le N(\lambda_M + N\delta_1)^k||\mathbf{x}(0)|| + \delta_2 N \frac{1-(\lambda_M + N\delta_1)^k}{1-(\lambda_M + N\delta_1)} \tag{5.40}$$

The arbitrarily assigned eigenvalue λ_M can be always selected such that $\lambda_M + \frac{||\Delta||N\bar{\lambda}}{\max\{\varepsilon_i\}} < 1$, where $\bar{\lambda}$ and ε_i are chosen compatibly with the procedure described in the previous section and such that $\frac{||\Delta||N\bar{\lambda}}{\max\{\varepsilon_i\}} < 1$. It follows that the state dynamics are ultimately bounded according to

$$\lim_{k\to\infty} ||\mathbf{x}(k)|| \le \frac{\delta_2 N}{1-(\lambda_M + N\delta_1)} \tag{5.41}$$

◇

The proposed control algorithm features the possibility of improving the time domain performance of the closed loop system. Indeed, the extra parameters contained in the time-varying terms of the sliding surface can enhance the transient shaping capabilities inherent to sliding mode based controllers. The following result proves that the discussed controller, beside ensuring robustness to bounded matched uncertainties, allows also performance requirements to be satisfied. It also gives clues on the correct choice of the tuning parameters.

Corollary 5.1. *Consider the plant (5.6) subject to the saturation constraint (4.41) and fed by the control input (5.9). The following shaping requirement on the transient response* $\mathbf{x}_t(k)$*:*

$$||\mathbf{x}_t(k)|| \le C_t\eta^k$$

can be always fulfilled for arbitrary $C_t, \eta \in \mathbb{R}$ *suitably choosing the parameters of the surface (5.8).*

Proof. From the previous Theorem one has

$$||\mathbf{x}(k)|| \leq \left[N||\mathbf{x}(0)|| - \frac{N\delta_2}{1-\lambda_2}\right]\lambda_2^k + \frac{N\delta_2}{1-\lambda_2} \tag{5.42}$$

with $\lambda_2 = \lambda_M + N\delta_1$ (with reference to quantities defined in the proof of the previous theorem). Setting $\lambda_2 = \eta$ and assigning $\tilde{\mathbf{A}}$ such that N satisfies: $N\left[||\mathbf{x}(0)|| - \frac{\delta_2}{1-\lambda_2}\right] = C_t$, the statement follows. ◇

Remark 5.2. *Differently from static state feedback control laws, the time-varying controller presented allows transient characteristics to be assigned by means of the time-varying part of the control law. This is clearly visible in Corollary 5.1, where it is explicitly given the bound of the state and is shown that it depends on $\bar{\lambda}$. Moreover, the proof of Corollary 5.1 shows the dependence of this bound from initial conditions.*

Remark 5.3. *The proofs of Theorem 5.3 and Corollary 5.1 provide an operative way of choosing "good" controller parameters. In fact, defining $\delta_1 = \frac{||\Delta||\bar{\lambda}}{\max\{\varepsilon_i\}}$, λ_M as the largest eigenvalue of $\tilde{A}$, and choosing $\bar{\lambda}$ such that the bound (5.31) and the following inequality:*

$$\lambda_M + N\delta_1 < 1$$

are satisfied, it can be shown that:

$$||x(k)|| \leq \left[N||x(0)|| - \frac{N\delta_2}{1-\lambda_2}\right]\lambda_2^k + \frac{N\delta_2}{1-\lambda_2}$$

being $\lambda_2 = \lambda_M + N\delta_1$ and δ_2 a positive constant (see proof of Theorem 5.2). Therefore, transient shaping can be obtained, suitably assigning the values of λ_2 and of $\left[N||x(0)|| - \frac{N\delta_2}{1-\lambda_2}\right]$ by means of matrix C. Note that the quantity $\left[N||x(0)|| - \frac{N\delta_2}{1-\lambda_2}\right]$ depends on the initial state, as expected.

5.2.5 A Benchmark Test

Consider the example (without uncertainties) treated in [11] and improved in [60], described by:

$$\mathbf{A} = \begin{bmatrix} 0 & 1 & 0 & 0 \\ 0 & 0 & 1 & 0 \\ 0 & 0 & 0 & 1 \\ -1 & 2\sqrt{2} & -4 & 2\sqrt{2} \end{bmatrix} \quad \mathbf{B} = \begin{bmatrix} 0 & 0 & 0 & 1 \end{bmatrix}^T. \tag{5.43}$$

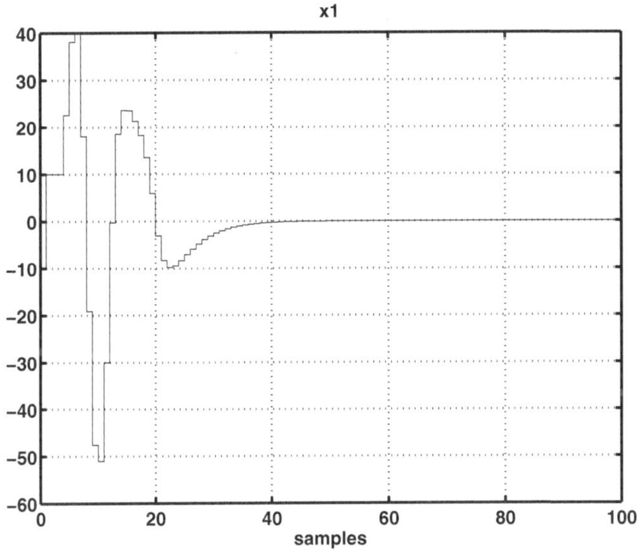

Fig. 5.1. State variable $x_1(k)$: plant (5.43) driven by controller (5.9)

with saturation threshold $M = 4$ initial conditions $\mathbf{x}(0) = [-10\ 10\ 10\ 10]^T$. Setting the proposed algorithm as follows: $\bar{\lambda} = 0.05$, $\mathbf{C} = \left[-0.08\ 0.66\ -1.5\ 1\right]$, $\mathbf{D} = \left[-20\ 100\ -5\ 0\right]$, the results reported in Figs. 5.1-5.2 have been obtained, showing the time response of the state variable $x_1(k)$ and the control signal $v(k)$ respectively. Simulations prove that a remarkable improvement of the transient response with respect to [11], [60] is achieved, both for response amplitude (peak of nearly 700 in [60] vs. 50 in our case) and transient duration (nearly 150 samples in [60] vs. 40 samples in our case). This is due to the transient shaping capability offered by the time-varying sliding surface.

5.3 Experimental Data: Stabilization of a Twin Rotor System

In this section, a further discrete-time control law is proposed along with experimental tests on a twin-rotor system. The controller is based on a time-varying sliding surface, different from that one proposed in the previous section, and it can be shown to be able to provide finite time plant stabilization for completely known systems. An extension to the case when matched bounded uncertainties affect the plant will be also considered, and a discrete-time sliding mode controller will be next proposed ensuring ultimate boundedness of state trajectories. Finally, experimental tests will be proposed relative to a twin-rotor system, in order to provide an experimental validation of the controller.

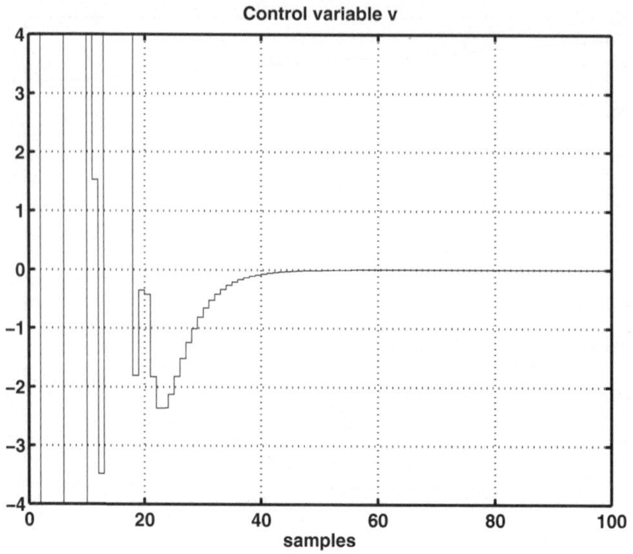

Fig. 5.2. Control input $v(k)$: plant (5.43) driven by controller (5.9)

5.3.1 Problem Statement

Consider the following discrete-time, time invariant SISO plant $S \stackrel{\text{def}}{=} \{\hat{\mathbf{A}}, \hat{\mathbf{B}}\}$ described by:

$$\hat{\mathbf{x}}(k+1) = \hat{\mathbf{A}}\hat{\mathbf{x}}(k) + \hat{\mathbf{B}}u(k) \tag{5.44}$$

where: $\hat{\mathbf{x}}(k) = [\hat{x}_1(k) \cdots \hat{x}_n(k)]^T \in \mathbb{R}^n$ is the state vector (assumed available for measurement), $u(k) \in \mathbb{R}$ is the control input, and $\hat{\mathbf{A}} \in \mathbb{R}^{n \times n}$, $\hat{\mathbf{B}} \in \mathbb{R}^n$ are the state and input distribution matrices, respectively.

Assumption 5.2. *The plant is controllable.*

The plant is supposed to be preceded by a saturating device $u(k) = \text{sat}_M(v(k))$ (2.2) with threshold M known.

Under the controllability hypothesis, there exists a smooth change of coordinates: $\mathbf{x}(k) = \mathbf{T}_2\mathbf{T}_1\hat{\mathbf{x}}(k)$ such that by $\mathbf{T}_1$ the plant is transformed in the controllability form, and $\mathbf{T}_2$ is such that system (5.44) becomes:

$$\mathbf{x}(k+1) = \mathbf{A}\mathbf{x}(k) + \mathbf{B}u(k) \tag{5.45}$$

with:

$$\mathbf{A} = \mathbf{T}_2\mathbf{T}_1\hat{\mathbf{A}}\mathbf{T}_1^{-1}\mathbf{T}_2^{-1} = \begin{bmatrix} \mathbf{A}_{11} & \mathbf{A}_{12} \\ \mathbf{A}_{21} & A_{22} \end{bmatrix} = \\ = \begin{bmatrix} 0 & \alpha_1 & 0 & \dots & 0 \\ 0 & 0 & \alpha_2 & \dots & 0 \\ \vdots & & \dots & & \alpha_{n-1} \\ a_1 & a_2 & a_3 & \dots & a_n \end{bmatrix}, \tag{5.46}$$

$$\mathbf{B} = \mathbf{T}_2\mathbf{T}_1\hat{\mathbf{B}} = \begin{bmatrix} \mathbf{0} \\ b \end{bmatrix} \quad |\alpha_1| < 1; \dots |\alpha_{n-1}| < 1 \tag{5.47}$$

Since $|\alpha_1| < 1; \dots |\alpha_{n-1}| < 1$, it is straightforward that, when all the eigenvalues of $\mathbf{A}$ are equal to zero, i.e. when $a_1 = a_2 = \dots = a_n = 0$, matrix $\mathbf{A}$ has norm less than 1.

5.3.2 A Finite Time Stabilizing Controller with Saturating Inputs

The basic idea pursued in this section is to design a time-varying sliding surface such that the achievement of a quasi sliding motion on it can be ensured with saturating input. The associated sliding mode based controller is used to drive the plant state toward a suitable neighborhood of the origin, where a standard state feedback controller can be used to achieve finite time stability.

First of all, a set of initial states can be easily found, starting from which the state vector can be directly steered to the origin using a standard linear state feedback controller. To this purpose, with reference to the transformed plant (5.45), consider the following control law:

$$u(k) = \mathbf{K}\mathbf{x}(k) \tag{5.48}$$

where $\mathbf{K}$ is such that the matrix $\mathbf{A} + \mathbf{B}\mathbf{K} \stackrel{\text{def}}{=} \mathbf{N}$ is nilpotent. In the following, the symbol $||\cdot||$ will denote $||\cdot||_2$.

Lemma 5.2. *It is given the discrete-time system (5.45) preceded by the saturating device (2.2) under Assumption 5.2. The deadbeat controller (5.48) guarantees finite time stabilization with saturating inputs for any initial condition belonging to the set:*

$$\mathscr{I} = \left\{ \mathbf{x}(0) : ||\mathbf{x}(0)|| \le \frac{M}{||\mathbf{K}||} \stackrel{\text{def}}{=} \bar{M} \right\} \tag{5.49}$$

Such set is an invariant set.

Proof. Consider $\mathbf{x}(0)$ as the initial condition, and apply the deadbeat controller $u(k) = \mathbf{K}\mathbf{x}(k)$. The saturation constraint provides:

$$|\mathbf{K}\mathbf{x}(k)| \le M \tag{5.50}$$

Moreover, the following chain of inequalities is straightforward:

$$|\mathbf{Kx}(k)| \leq ||\mathbf{K}|| \cdot ||\mathbf{x}(k)|| \leq ||\mathbf{K}|| \cdot ||\mathbf{N}^k|| \cdot ||\mathbf{x}(0)|| \\ \leq ||\mathbf{K}|| \cdot ||\mathbf{N}||^k \cdot ||\mathbf{x}(0)|| \leq ||\mathbf{K}|| \cdot ||\mathbf{x}(0)|| \tag{5.51}$$

since $||\mathbf{N}|| < 1$. In fact, $\mathbf{N} = \mathbf{A} + \mathbf{BK}$ is nilpotent and therefore all the elements of its last row are equal to zero, due to transformation matrix $\mathbf{T}_1$: as a consequence $||\mathbf{N}|| < 1$, due to transformation matrix $\mathbf{T}_2$, i.e. to the choice of coefficients $\alpha_1 < 1 \ldots \alpha_n < 1$. Expression (5.51) implies that:

$$||\mathbf{x}(0)|| \leq \frac{M}{||\mathbf{K}||} \Rightarrow ||\mathbf{x}(k)|| \leq \frac{M}{||\mathbf{K}||}$$

i.e. the set $\mathscr{I}$ is invariant. In other words, if the initial state belongs to the set $\mathscr{I}$ and fulfills the saturation constraints, the entire dynamics satisfy the same constraint. Moreover, the deadbeat controller ensures stabilization in finite time. ♢

As already mentioned, a time varying sliding surface will be introduced. As well known [77] [78], a vector $\mathbf{C} = [\mathbf{C}_1\ C_2] \in \mathbb{R}^n$ can be chosen such that, when a sliding motion is achieved on the following sliding surface:

$$\hat{s}(k) = \mathbf{Cx}(k) = \mathbf{C}_1\mathbf{x}_1(k) + C_2x_2(k) = 0 \tag{5.52}$$

the corresponding reduced order system has assigned stable eigenvalues, and, as a consequence, system (5.45) is stable, too. It will be assumed here to choose $\mathbf{C}_1$ and C_2 such that the matrix

$$\mathbf{N}_1 = \mathbf{A}_{11} - \mathbf{A}_{12}\frac{\mathbf{C}_1}{C_2}$$

has stable eigenvalues, and that, without loss of generality, $C_2 > 0$. Starting from the classical sliding surface (5.52), always with reference to the transformed plant (5.45), the following time varying sliding surface can be introduced:

$$s(k) = \mathbf{Cx}(k) - \lambda^k\mathbf{CAx}(k-1) = 0 \tag{5.53}$$

where $0 < \lambda < 1$ is a design parameter. It is straightforward that when $s(k) = 0$, the system is asymptotically stable, since for $k \to \infty$ surface (5.53) tends to surface (5.52). The equivalent control ensuring the achievement of a sliding motion on (5.53) can be obtained imposing the condition $s(k+1) = 0$, i.e.:

$$u_{eq}(k) = -(\mathbf{CB})^{-1}\mathbf{CAx}(k)(1 - \lambda^{k+1}) \tag{5.54}$$

After the application of the controller (5.54), the closed loop system for $k \geq 1$ is described by:

$$\mathbf{x}(k+1) = \left[\mathbf{A} - (1 - \lambda^{k+1})\mathbf{B}(\mathbf{CB})^{-1}\mathbf{CA}\right]\mathbf{x}(k) \\ = \mathbf{F}(\lambda, k+1)\mathbf{x}(k) \tag{5.55}$$

Due to the choice of **C** and λ in (5.53), the closed loop system (5.55) is asymptotically stable. The following result can be proved:

Theorem 5.4. *It is given the discrete-time system (5.45) preceded by the saturating device (2.2) under Assumption 5.2. The controller (5.54) guarantees that any initial condition belonging to the set:*

$$\mathscr{J} = \left\{ \mathbf{x}(0) \in \left(\prod_{j=0}^{k-1} \mathbf{Q}(\lambda, k-j) \right) \mathscr{I} \right\} \tag{5.56}$$

being $\mathbf{Q}(\lambda, k-j) = \mathbf{F}(\lambda, k-j)^{-1}$, *is driven to the set* $\mathscr{I}$ *in k steps without violating the saturation constraints. Therefore, finite time stabilization with saturating inputs is guaranteed for any initial condition belonging to the set* $\mathscr{J}$ *by coupling (5.48) and (5.54).*

Proof. The proof consists in showing that a procedure exists for selecting the parameter λ such that the control law (5.54) drives the state into the invariant set $\mathscr{I}$ in a finite number of steps k. Let's consider $\overline{\mathbf{x}} \in \partial\mathscr{I}$, hence $||\overline{\mathbf{x}}|| = \overline{M}$.

- *Step* 1 Consider the initial state $\hat{\mathbf{x}}_1$

  $$\hat{\mathbf{x}}_1 := \mathbf{Q}(\lambda, 1)\overline{\mathbf{x}} := [\mathbf{A} - (1-\lambda)\mathbf{B}(\mathbf{CB})^{-1}\mathbf{CA}]^{-1}\overline{\mathbf{x}}.$$

  Imposing that $|u_{eq}(0)| \leq M$, one gets

  $$||(1-\lambda)\mathbf{Q}(\lambda,1)|| \leq \frac{||\mathbf{K}||}{||(\mathbf{CB})^{-1}\mathbf{CA}||} = \frac{M}{\overline{M}||(\mathbf{CB})^{-1}\mathbf{CA}||} \tag{5.57}$$

  Denoting by

  $$\lambda_1 = \inf\{\lambda \in (0,1) : (5.57) \text{ is fulfilled}\},$$

  one immediately gets that initial conditions belonging to the set $\mathbf{Q}(\lambda,1)\mathscr{I}$ can be driven to the set $\mathscr{I}$ simply setting $1 > \lambda > \lambda_1$ in (5.54), in fact

  $$\begin{aligned} \mathbf{x}(1) &= \mathbf{A}\hat{\mathbf{x}}_1 + \mathbf{B}u_{eq}(0) = \mathbf{F}(\lambda,1)\hat{\mathbf{x}}_1 = \\ &= \mathbf{F}(\lambda,1)\mathbf{Q}(\lambda,1)\overline{\mathbf{x}} = \overline{\mathbf{x}}. \end{aligned} \tag{5.58}$$

- *Step* 2 Define $\hat{\mathbf{x}}_2$ as

  $$\hat{\mathbf{x}}_2 := \mathbf{Q}(\lambda,1)\mathbf{Q}(\lambda,2)\overline{\mathbf{x}}$$

  The saturation constraints is fulfilled if the parameter λ in (5.54) is chosen as $1 > \lambda > \lambda_2$, with λ_2 solution of

  1. $||(1-\lambda)\mathbf{Q}(\lambda,1)\mathbf{Q}(\lambda,2)|| < \mu$,
  2. $||(1-\lambda^2)\mathbf{Q}(\lambda,2)|| < \mu$,

  having defined $\mu = \frac{||\mathbf{K}||}{||(\mathbf{CB})^{-1}\mathbf{CA}||}$. It follows that all initial conditions belonging to $\mathbf{Q}(\lambda,1)\mathbf{Q}(\lambda,2)\mathscr{I}$ can be driven to the set $\mathscr{I}$ in 2 steps.

- *Step k* As before define

$$\hat{\mathbf{x}}_k := \left(\prod_{j=0}^{k-1}\mathbf{Q}(\lambda, k-j)\right)\bar{\mathbf{x}}$$

and imposing that

$$\left\|(1-\lambda^r)\prod_{j=0}^{k-r}\mathbf{Q}(\lambda, k-j)\right\| < \mu \quad \forall r \leq k \tag{5.59}$$

one gets that, denoting

$$\lambda_k := \inf\{\lambda \in (0,1) : (5.59) \text{ is fulfilled}\},$$

and choosing $1 > \lambda > \lambda_k$ in (5.54), any initial condition belonging to the set $\mathscr{J}$ can be driven to the set $\mathscr{I}$ in k steps. Therefore the statement follows. $\diamondsuit$

5.3.3 Presence of Bounded Uncertainties

The case when matched bounded disturbances or uncertainties affect the plant will be now considered. As well known, such class of disturbances is traditionally dealt with by sliding mode control, though may be restrictive for some plants. Reference is made here to the plant:

$$\begin{aligned} \mathbf{x}(k+1) &= \mathbf{A}\mathbf{x}(k) + \mathbf{B}[u(k)+d(k)] = \\ &= \begin{bmatrix} \mathbf{A}_{11} & \mathbf{A}_{12} \\ \mathbf{A}_{21} & A_{22} \end{bmatrix}\mathbf{x}(k) + \begin{bmatrix} \mathbf{0} \\ b \end{bmatrix}[u(k)+d(k)] \end{aligned} \tag{5.60}$$

under the following assumption:

Assumption 5.3. *The uncertain term $d(k)$ is such that: $|d(k)| \leq \rho$, being ρ a known constant. Moreover, ρ is such that*

$$\rho < \frac{M}{n||\mathbf{B}|| \cdot ||\mathbf{K}||}.$$

It is straightforward to verify that Lemma 5.2 can be extended to cope with the uncertain plant (5.60) as follows:

Lemma 5.3. *It is given the discrete-time system (5.60) preceded by the saturating device (2.2) under Assumptions 5.2, 5.3. The deadbeat controller (5.48) guarantees that for any initial condition belonging to the set:*

$$\mathscr{I}_\rho = \left\{ \mathbf{x}(0) : ||\mathbf{x}(0)|| \leq \frac{M}{||\mathbf{K}||} - L \stackrel{\text{def}}{=} M^* \right\} \tag{5.61}$$

with $L = n\rho||\mathbf{B}||$*, ultimate boundedness of the state trajectories is ensured according to*

$$\lim_{k\to\infty} ||\mathbf{x}(k)|| \leq L \tag{5.62}$$

Following the lines of the previous section, the following result can be stated.

Theorem 5.5. *It is given the discrete-time system (5.60) preceded by the saturating device (2.2) under Assumptions 5.2, 5.3. For perturbing terms* $d(k)$ *bounded by a constant satisfying:*

$$\rho < \min\left\{ \frac{M}{2n||\mathbf{K}|| \cdot ||\mathbf{B}||}, \frac{M}{n\rho||\mathbf{A}||} \right\}. \tag{5.63}$$

the control law $u(k) = u^{eq}(k) + v(k)$*, with* $u^{eq}(k)$ *given by (5.54) and* $v(k)$ *of the form*

$$v(k) = \tag{5.64}$$

$$\begin{cases} 0 & \text{if } k = 0 \\ -\bar{\mathbf{D}}\mathbf{F}(\lambda, k+1)(\mathbf{x}(k) - \prod_{j=1}^{k} \mathbf{F}(\lambda, j)\mathbf{x}(0)) & \text{if } k \geq 1 \end{cases}$$

with $\bar{\mathbf{D}} = [0\ 0\ \cdots\ 0\ \frac{1}{b}]$*, guarantees that any initial condition belonging to the set:*

$$\mathscr{J}_\rho = \left\{ \mathbf{x}(0) \in \left(\prod_{j=0}^{k-1} \mathbf{Q}(\lambda, k-j) \right) \mathscr{I}_\rho \right\} \tag{5.65}$$

is driven to the set $\mathscr{I}_\rho$ *in* k *steps without violating the saturation constraints. Therefore, ultimate boundedness of state trajectories is guaranteed for any initial condition belonging to the set* $\mathscr{J}_\rho$ *by coupling (5.48), (5.54) and (5.64).*

Proof. The proof consists in showing that a procedure exists for selecting the parameter λ such that the control laws (5.54) and (5.64) drive the state into the invariant set $\mathscr{I}_\rho$ in a finite number of steps k. Noticing that

$$(\mathbf{I} - \mathbf{B}\bar{\mathbf{D}})\mathbf{F}(\lambda, 1) = (\mathbf{I} - \mathbf{B}\bar{\mathbf{D}})\mathbf{A} = \mathbf{N},$$

since $\mathbf{N}^n = 0$, the following expressions can be easily derived:

$$
\begin{aligned}
\mathbf{x}(1) &= \mathbf{F}(\lambda,1)\mathbf{x}(0) + \mathbf{B}d(0)\\
\mathbf{x}(2) &= \mathbf{F}(\lambda,2)\left(\mathbf{F}(\lambda,1)\mathbf{x}(0)) + \mathbf{B}d(0)\right) + \mathbf{B}v(1) + \mathbf{B}d(1)\\
&= \mathbf{F}(\lambda,2)\left(\mathbf{F}(\lambda,1)\mathbf{x}(0) + \mathbf{B}d(0)\right)\\
&- \mathbf{B}\bar{\mathbf{D}}\mathbf{F}(\lambda,2)\mathbf{B}d(0) + \mathbf{B}d(1)\\
&= \mathbf{F}(\lambda,2)\mathbf{F}(\lambda,1)\mathbf{x}(0) + \mathbf{N}\mathbf{B}d(0) + \mathbf{B}d(1)
\end{aligned}
$$

$$
\begin{aligned}
\mathbf{x}(3) &= \mathbf{F}(\lambda,3)\mathbf{F}(\lambda,2)\mathbf{F}(\lambda,1)\mathbf{x}(0) + \mathbf{F}(\lambda,3)(\mathbf{B}d(1)\\
&+ \mathbf{N}\mathbf{B}d(0)) + \mathbf{B}d(2) + \mathbf{B}v(2)\\
&= \mathbf{F}(\lambda,3)\mathbf{F}(\lambda,2)\mathbf{F}(\lambda)\mathbf{x}(0) + \mathbf{B}d(2)\\
&+ (\mathbf{I} - \mathbf{B}\bar{\mathbf{D}})\mathbf{F}(\lambda,3)(\mathbf{B}d(1) + \mathbf{N}\mathbf{B}d(0))\\
&= \mathbf{F}(\lambda,3)\mathbf{F}(\lambda,2)\mathbf{F}(\lambda)\mathbf{x}(0) + \mathbf{B}d(2)\\
&+ \mathbf{N}\mathbf{B}d(1) + \mathbf{N}^2\mathbf{B}d(0)\\
&\vdots\\
\mathbf{x}(k) &= \prod_{j=1}^{k} \mathbf{F}(\lambda,j)\mathbf{x}(0) + \sum_{i=0}^{n-1} \mathbf{N}^i\mathbf{B}d(k-i)
\end{aligned}
$$

Moreover, the control input (5.64) fulfills the following inequality

$$
\begin{aligned}
|v(k)| &= \left|\bar{\mathbf{D}}\mathbf{F}(\lambda,k+1)\sum_{i=0}^{n-1}\mathbf{N}^i\mathbf{B}d(k-i-1)\right|\\
&\le \rho||\bar{\mathbf{D}}|| \cdot \sup_{\lambda\in(0,1)} ||\mathbf{F}(\lambda,1)|| \cdot \sum_{i=0}^{n-1} ||\mathbf{N}^i\mathbf{B}||\\
&\le n\rho||\bar{\mathbf{D}}|| \cdot ||\mathbf{B}|| \cdot ||\mathbf{A}|| = n\rho||\mathbf{A}||.
\end{aligned}
$$

and accounting for the saturation constraint requires that

$$
\rho < \frac{M}{n||A||}
$$

therefore the condition (5.63) is found coupling the previous inequality and Assumption 5.3. Following the same approach of the proof of Theorem 5.4, it is now enough to impose $\forall r \le k$

$$
\left\| (1-\lambda^r)\prod_{j=0}^{k-r}\mathbf{Q}(\lambda,k-j) \right\| < \frac{M - n\rho||\mathbf{A}||}{||(\mathbf{CB})^{-1}\mathbf{CA}||(M^* - n\rho||\mathbf{B}||)} \tag{5.66}
$$

Denoting by

$$\lambda_k = \inf\{\lambda \in (0,1) : \text{condition (5.66) is fulfilled}\}$$

one can conclude that choosing $1 > \lambda > \lambda_k$ in (5.54) any initial condition belonging to the set $\mathscr{J}_\rho$ can be driven to the set $\mathscr{I}_\rho$ in k steps using a control law satisfying $|u_{eq}(k)| \leq M - n\rho||\mathbf{A}|| < M$.

In fact, if $\bar{\mathbf{x}} \in \mathscr{I}_\rho$, then for the initial condition $\mathbf{x}(0) := \left(\prod_{j=0}^{k-1}\mathbf{Q}(\lambda, k-j)\right)\bar{\mathbf{x}}$ it holds

$$\begin{aligned}
\mathbf{x}(k) &= \prod_{j=1}^{k}\mathbf{F}(\lambda, j)\mathbf{x}(0) + \sum_{i=0}^{n-1}\mathbf{N}^i\mathbf{B}d(k-i) \\
&= \prod_{j=1}^{k}\mathbf{F}(\lambda, j)\left(\prod_{j=0}^{k-1}\mathbf{Q}(\lambda, k-j)\right)\bar{\mathbf{x}} + \sum_{i=0}^{n-1}\mathbf{N}^i\mathbf{B}d(k-i) \\
&= \bar{\mathbf{x}} + \sum_{i=0}^{n-1}\mathbf{N}^i\mathbf{B}d(k-i)
\end{aligned}$$

hence

$$||\mathbf{x}(k)|| \leq ||\bar{\mathbf{x}}|| + n\rho||B|| < M^* - n\rho||\mathbf{B}|| + n\rho||\mathbf{B}|| = M^*. \qquad \diamond$$

5.3.4 Experimental Results

Previous theoretical results have been experimentally validated on the twin rotor shown in Fig. 5.3. The plant has been built in our laboratory for educational purposes, and is constituted of two metal arms: the first is locked to the ground, while the second is linked to the first one, and can move with two degrees of freedom. The movements are generated by two brushless D.C. motors (produced by AIRPAX©), placed on the two ends of the free arm. Moreover, two potentiometers are in charge of measuring the angular displacements of the free arm.

The overall control system is depicted in Fig.5.4. The controller code is written in MATLAB/SIMULINK©, running on a Personal Computer (PC). The PC is equipped with a Plug-and-Play general purpose board, namely **NI-PCI6024e**, produced by NATIONAL INSTRUMENTS©, which is connected to MATLAB/ SIMULINK© by means of Real Time Workshop and Real Time Windows Target MATLAB© packages. The **NI-PCI6024e** allows data exchange between PC and the plant, but it is not directly connected to the potentiometers and to the motors. An interface board, made in our Lab, is in charge to filter and to adapt the following signals:

- signals coming from the sensors, before passing them to the **NI-PCI6024e** board, and therefore to PC;
- signals coming from the controller, i.e. from the PC through the **NI-PCI6024e** board, before passing them to the power board.

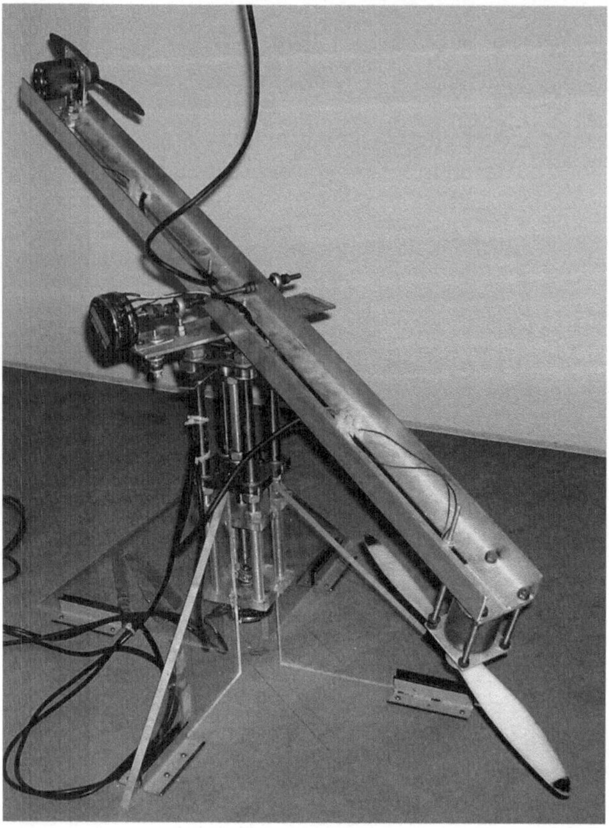

Fig. 5.3. The twin rotor

The power board, made in our Lab, too, mounts a PWM modulator and drives the motors with suitable voltages, corresponding to the control actions produced by the control law implemented in SIMULINK©. The maximum range that the power board can supply is $\pm 12V$. Nevertheless, the safer voltage saturation limit of $\pm 7V$ was chosen, because of the problems we encountered during the testing phase of the overall control system. Indeed sudden changes in the control variables caused damages to the boards (microchips and capacitors, for example), and, more seldom, a risk occurred to burn the motors. Finally, note that the angular velocities, required by the state feedback control law, were obtained by filtering and differentiating the signals coming from the potentiometer. The chosen sampling time was $T_c = 0.05\ s$.

The mathematical model of the plant can be derived using well known theoretical physics results. Making reference to Fig.5.5, and introducing the state vector $\mathbf{x} = \begin{bmatrix} x_1\ x_2\ x_3\ x_4 \end{bmatrix}^T = \begin{bmatrix} \theta\ \phi\ \dot{\theta}\ \dot{\phi} \end{bmatrix}^T$, where θ and ϕ are the pitch and yaw angles, respectively, the twin rotor is described by the following nonlinear equation $\dot{\mathbf{x}} = \mathbf{f}(\mathbf{x}) + \mathbf{h}(\mathbf{x})\mathbf{u}$, with:

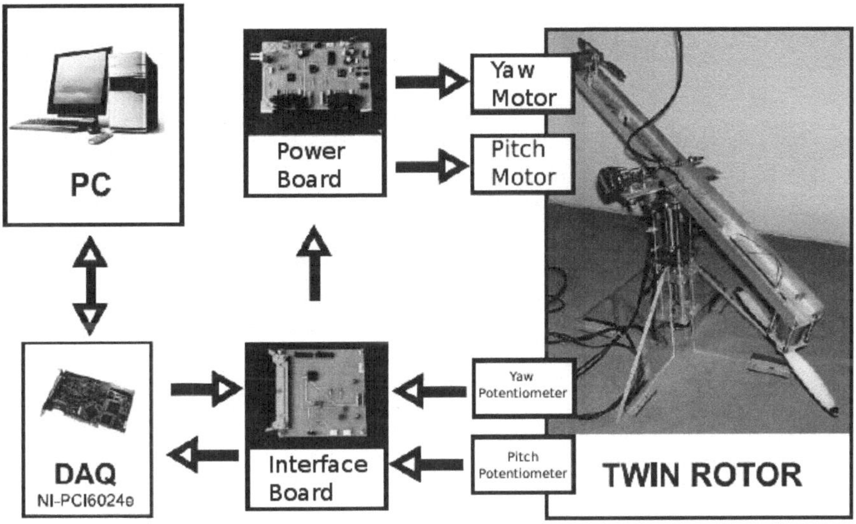

Fig. 5.4. The twin rotor control scheme

$$\mathbf{f}(\mathbf{x}) = \begin{bmatrix} x_3 \\ x_4 \\ \dfrac{-mgl_c \cos x_1 - J_L x_4^2 \cos x_1 \sin x_1 - \alpha_\theta x_3}{J_L} \\ \dfrac{2J_L x_3 x_4 \cos x_1 \sin x_1 - \alpha_\phi x_4}{(\cos^2 x_1 J_L + J_A)} \end{bmatrix} \tag{5.67}$$

$$\mathbf{h}(\mathbf{x}) = \begin{bmatrix} 0 & 0 \\ 0 & 0 \\ \dfrac{l_1 p_1}{J_L} & \dfrac{p_2}{J_L} \\ \dfrac{p_3 cos(x_1)}{(\cos^2 x_1 J_L + J_A)} & \dfrac{l_2 p_4 cos x_1}{(\cos^2 x_1 J_L + J_A)} \end{bmatrix} \tag{5.68}$$

where

- $\mathbf{u} = \begin{bmatrix} u_1 & u_2 \end{bmatrix}^T$ is the input vector, i.e. the voltages of the two motors driving the twin-rotor system;
- J_L and J_A are the inertia moments of the free arm and of locked one, respectively;
- l_c is the centre of gravity of the free arm;
- l_1 and l_2 are the distances between the ends and the centre of the free arm;
- g is the gravity acceleration;
- α_ϕ and α_θ are the damper coefficients for angles ϕ and θ, respectively;

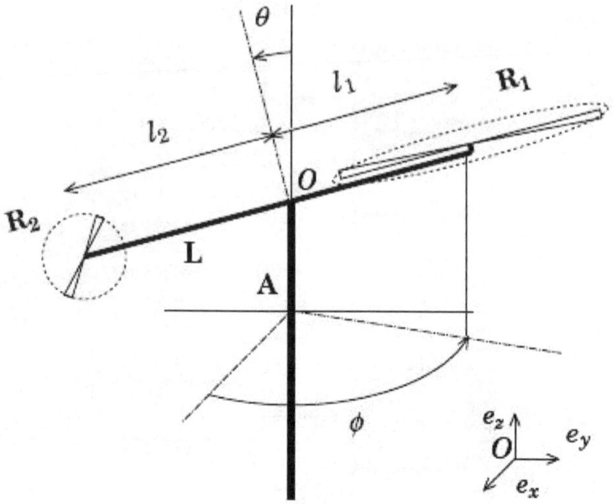

Fig. 5.5. The twin rotor model

- p_1, p_2, p_3, p_4 are suitable coefficients, correlating the voltages and the moments supplied by the motors.

The physical parameters of the plant, described in Fig.5.5, are reported in Tab.5.1. Note that all the numerical values have been found experimentally.

Table 5.1. Physical parameters of the twin rotor

Parameter	Value	Unit
J_L	0.1087	$[kg \cdot m^2]$
J_A	0	$[kg \cdot m^2]$
m	1.115	$[Kg]$
l_c	0.0033	$[m]$
l_1	0.375	$[m]$
l_2	0.375	$[m]$
g	9.81	$[m/s^2]$
α_ϕ	0.05	$[N \cdot m \cdot s]$
α_θ	0.05	$[N \cdot m \cdot s]$
p_1	0.08	$[kg \cdot m \cdot rad/(V \cdot s^2)]$
p_2	0.005	$[kg \cdot m^2 \cdot rad/(V \cdot s^2)]$
p_3	0.005	$[kg \cdot m^2 \cdot rad/(V \cdot s^2)]$
p_4	0.08	$[kg \cdot m \cdot rad/(V \cdot s^2)]$

The non-linear model has been linearized with respect to the equilibrium point $\left[\mathbf{x}_e^T \; \mathbf{u}_e^T\right]^T = \left[\mathbf{0}_{4x1}^T \; 1.24 \; -0.21\right]^T$, obtaining the following continuous time linear time invariant state space representation:

$$\frac{d}{dt}(\Delta\mathbf{x}) = \mathbf{A}_\Delta \Delta\mathbf{x} + \mathbf{B}_\Delta \Delta\mathbf{u}$$

being $\Delta\mathbf{u} = \mathbf{u} - \mathbf{u}_e$, $\Delta\mathbf{x} = \mathbf{x} - \mathbf{x}_e$, and with:

$$\mathbf{A}_\Delta = \begin{bmatrix} 0 & 0 & 1 & 0 \\ 0 & 0 & 0 & 1 \\ 0 & 0 & -0.46 & 0 \\ 0 & 0 & 0 & -0.46 \end{bmatrix} \quad \mathbf{B}_\Delta = \begin{bmatrix} 0 & 0 \\ 0 & 0 \\ 0.276 & 0.046 \\ 0.046 & 0.276 \end{bmatrix}$$

Considering θ and ϕ as the system output, i.e. $\mathbf{y} = \left[\phi \;\; \theta\right]^T$, the corresponding input-output transfer matrix is given by:

$$\mathbf{F}(s) = \begin{bmatrix} \dfrac{0.276}{s^2+0.46s} & \dfrac{0.046}{s^2+0.46s} \\ \dfrac{0.046}{s^2+0.46s} & \dfrac{0.276}{s^2+0.46s} \end{bmatrix}$$

Since $\mathbf{F}(s)$ is a diagonally dominant matrix, the coupling terms between θ and ϕ dynamics have been neglected in the linearised plant. The twin rotor has been considered as made of two independent SISO plants, characterized by the same transfer function $F_{11}(s) = F_{22}(s) = \dfrac{0.276}{s^2+0.46s}$. In order to apply the proposed control law, the above transfer functions have been discretized with a sampling time $T_c = 0.05\ s$, obtaining two equal subsystems of the form (5.44), with:

$$\hat{\mathbf{A}} = \begin{bmatrix} 1 & 0.0494 \\ 0 & 0.9773 \end{bmatrix}, \quad \hat{\mathbf{B}} = \begin{bmatrix} 0.0003 \\ 0.0136 \end{bmatrix}.$$

Successively, $\hat{\mathbf{A}}$ and $\hat{\mathbf{B}}$ have been transformed as in (5.45), (5.46), (5.47), obtaining:

$$\mathbf{A} = \begin{bmatrix} 0 & 0.8 \\ -1.22 & 1.98 \end{bmatrix}, \quad \mathbf{B} = \begin{bmatrix} 0 \\ 1.25 \end{bmatrix}.$$

Finally, two controllers have been built according to the approach described in Theorem 5.4, with $\mathbf{K} = \left[0.98 \; -1.58\right]$, $\mathbf{C} = \left[1 \; 1.2\right]$ and $\lambda = 0.999988$. The saturation threshold of actuators is $M = 7\,V$, as explained at the beginning of this section.

With reference to the theoretical development presented in Sections 5.3.2, the sets of initial conditions from which the state can be steered to the set $\mathscr{I}$ is reported in Fig.5.6. Accordingly, the initial conditions have been chosen as -31 *deg* for the pitch angle θ and 24 *deg* for the yaw angle ϕ, with null initial velocities, in order to drive the state to the set $\mathscr{I}$ in just one sampling time.

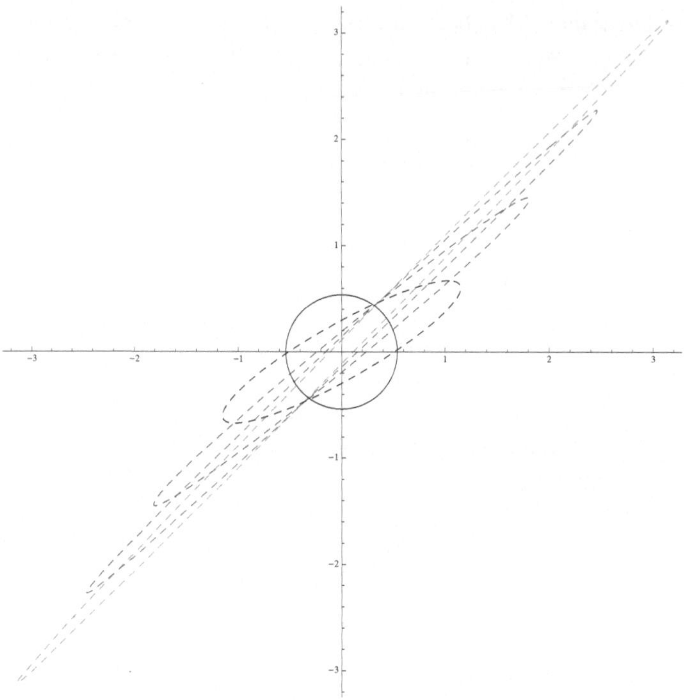

Fig. 5.6. Sets of initials conditions from which the state can be steered to the set $\mathscr{I}$ (the circle) in 1 (black), 2 (red), 3 (purple), 4 (green) steps

Two experiments have been performed. In the first one, the control law based on (5.53), and given by (5.54), (5.48), has been implemented for each subsystem. In the second one, a standard equivalent control law based on (5.52) has been implemented for each subsystem, i.e.

$$u_{eq}(k) = -(\mathbf{CB})^{-1}\mathbf{CAx}(k) \tag{5.69}$$

Results of the first experiment have been reported in Figs.5.7-5.8, showing the experimental yaw and pitch angles, and Figs.5.9-5.10 displaying the control inputs u_1 and u_2. The corresponding variables for the second experiment are reported in Figs.5.11-5.14. It can be noticed that when using control law (5.54), (5.48), the initial value of the control variables is always 0, regardless of the initial state, while using control law (5.69) the initial control effort depends on the initial state (the farther the initial state is from the origin, the larger will be the initial control effort). This fact can be seen comparing Figs.5.9-5.10 with Figs.5.13-5.14. However, after few time instants, control variables produced by (5.69) assume values comparable with signal produced by control law (5.54), (5.48). Anyway, the smaller initial control activity produced by (5.54), (5.48) is paid by the presence of some overshoots in the case of the behavior of the pitch angle, arising when controlled by the same controller (5.54), (5.48).

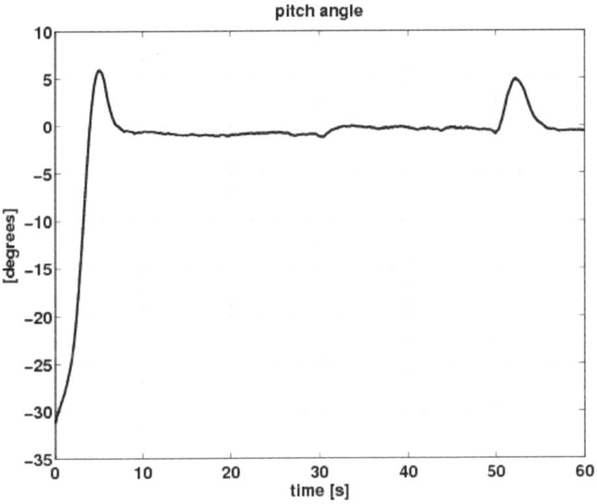

Fig. 5.7. Pitch angle θ (control law (5.54), (5.48))

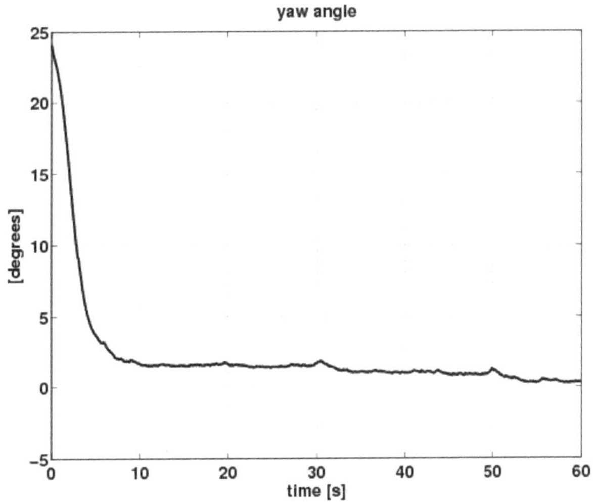

Fig. 5.8. Yaw angle ϕ (control law (5.54), (5.48))

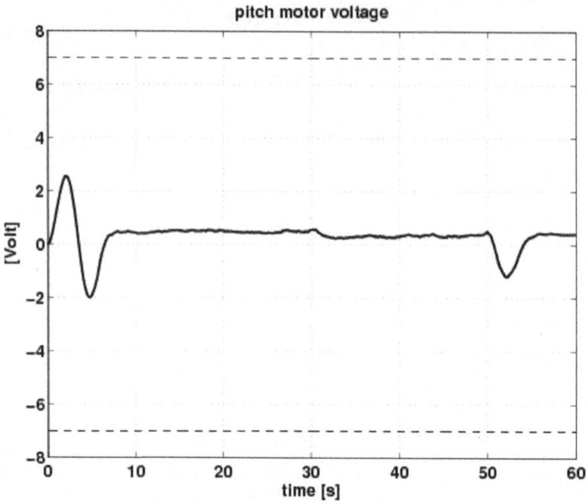

Fig. 5.9. Control input u_1 (control law (5.54), (5.48))

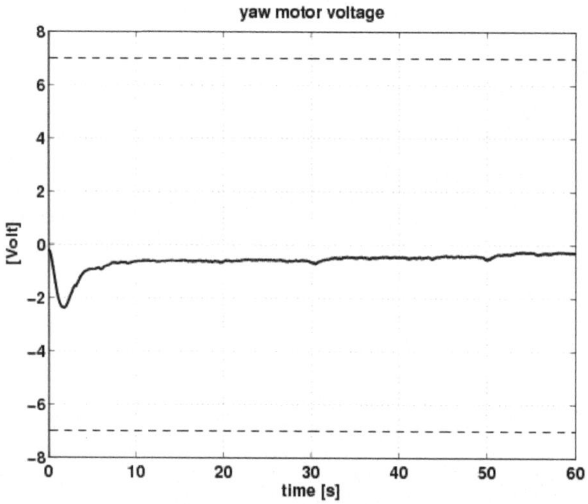

Fig. 5.10. Control input u_2 (control law (5.54), (5.48))

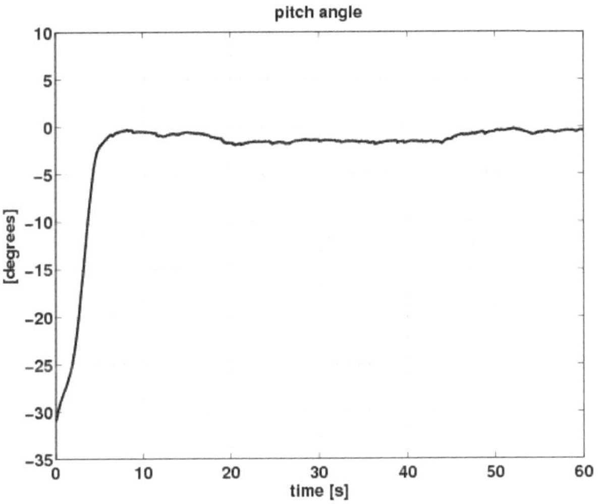

Fig. 5.11. Pitch angle θ (control law (5.69))

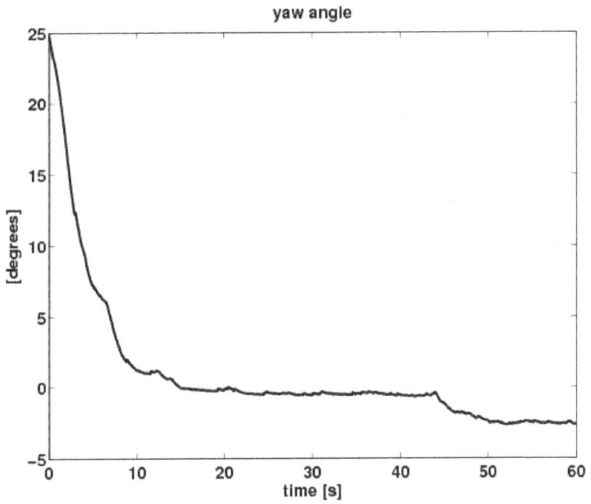

Fig. 5.12. Yaw angle ϕ (control law (5.69))

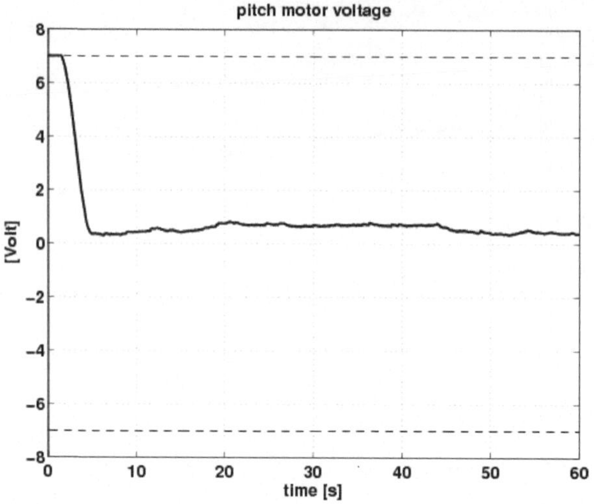

Fig. 5.13. Control input u_1 (control law (5.69))

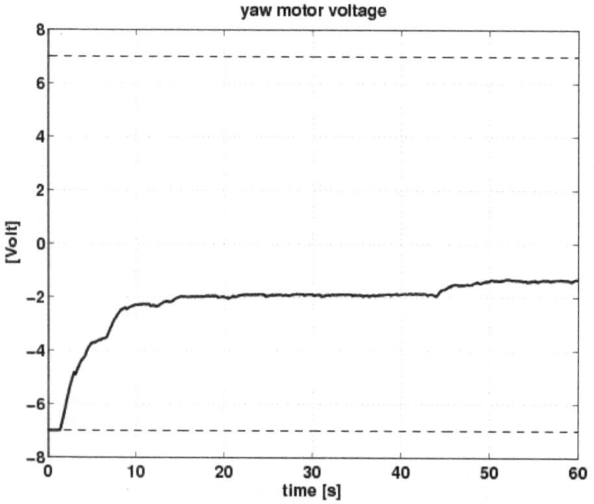

Fig. 5.14. Control input u_2 (control law (5.69))

Appendix A
Support Material

A.1 Support Material for Chapters 2, 3

This section contains some results from convex analysis and measure theory which have been employed for the characterization of null controllable regions in Chapter 2 and Chapter 3. The interested readers may refer to the textbooks [33], [34] and [79] for the proofs of the presented results and for further details.

A.1.1 Tools from Convex Analysis

We recall that, given k elements $\mathbf{x}_1, \mathbf{x}_2, ..., \mathbf{x}_k \in \mathbb{R}^n$, the vector sum

$$\lambda_1 \mathbf{x}_1 + \lambda_2 \mathbf{x}_2 + \cdots + \lambda_k \mathbf{x}_k$$

is called a *convex combination* if the coefficients λ_i are non-negative $\forall i$ and satisfy

$$\sum_{i=1}^{k} \lambda_i = 1.$$

The following simple proposition states a fundamental characterization of convex sets.

Proposition A.1. *A subset of $\mathbb{R}^n$ is convex if and only if it contains all the convex combinations of its elements.*

The convexity property is invariant under linear operations, namely

- if C is convex, so it is every translated set $C + \gamma$, $\gamma \in \mathbb{R}$;
- if C is convex, the for any $\alpha \in \mathbb{R}$ the rescaled set $\alpha C = \{\alpha \mathbf{x} : \mathbf{x} \in C\}$ is convex too;
- if C_1, C_2 are convex sets, then so is their sum $C_1 + C_2$ where

$$C_1 + C_2 = \{\mathbf{z} = \mathbf{x} + \mathbf{y} : \mathbf{x} \in C_1, \mathbf{y} \in C_2\}.$$

Moreover the following proposition can be easily proved.

Proposition A.2. *Let C_1 and C_2 be convex sets in $\mathbb{R}^m$ and $\mathbb{R}^p$, respectively. Then*

$$C_1 \times C_2 = \{\mathbf{z} = (\mathbf{x}, \mathbf{y}) : \mathbf{x} \in C_1, \mathbf{y} \in C_2\}$$

is a convex set in $\mathbb{R}^{m+p}$.

Given a set $S \in \mathbb{R}^n$ there exists a unique smallest affine set A containing S; this is called the *affine hull* of the set S and it is denoted by $A = \mathrm{aff}(S)$. The dimension of a convex set is defined as the dimension of its affine hull.

Let us denote by $\mathscr{D}$ the n-dimensional closed unit ball

$$\mathscr{D} := \{\mathbf{z} \in \mathbb{R}^n : ||\mathbf{z}|| \leq 1\}.$$

Given an arbitrary point $\mathbf{x} \in \mathbb{R}^n$ and a non empty set $\mathscr{A} \subset \mathbb{R}^n$, for $\eta > 0$ we define the sets

$$\mathbf{x} + \eta\mathscr{D} := \{\mathbf{z} \in \mathbb{R}^n : ||\mathbf{z} - \mathbf{x}|| \leq \eta\};$$

$$\mathscr{A} + \eta\mathscr{D} := \{\mathbf{z} \in \mathbb{R}^n : \exists\, \mathbf{a} \in \mathscr{A} : ||\mathbf{z} - \mathbf{a}|| \leq \eta\} = \bigcup_{\mathbf{a} \in \mathscr{A}} [\mathbf{a} + \eta\mathscr{D}].$$

The *closure* $\overline{C}$ and the *interior part* $\mathrm{int}(C)$ of a set $C \subset \mathbb{R}^n$ can be expressed in terms of the sets defined above:

$$\overline{C} = \bigcap_{\eta > 0} [C + \eta\mathscr{D}]$$

$$\mathrm{int}(C) = \{\mathbf{x} \in C : \exists\, \eta > 0 : [\mathbf{x} + \eta\mathscr{D}] \subset C\}$$

The *relative interior* part of a convex set C, $\mathrm{ri}(C)$, is defined as the interior part of the set C when it is regarded as a subset of its affine hull $\mathrm{aff}(C)$. Denoting by $\overline{C}$ the closure of C, the set of inclusions

$$\mathrm{ri}(C) \subset C \subset \overline{C}$$

is straightforward to verify.

Remark A.1. *If $C \subset \mathbb{R}^n$ is a n-dimensional convex set, i.e. if $\mathrm{aff}(C) = \mathbb{R}^n$, then the relative interior part of C coincides with the standard interior part $\mathrm{int}(C)$.*

Theorem A.1. *Let C be a convex set in $\mathbb{R}^n$; let $\mathbf{x} \in \mathrm{ri}(C)$ and $\mathbf{y} \in \overline{C}$. Then*

$$\theta\mathbf{x} + (1 - \theta)\mathbf{y} \in \mathrm{ri}(C)$$

for any $0 < \theta \leq 1$.

Proof. The proof is given for the case of a n-dimensional convex set; the general case can be deduced by the continuity of the projection operator. We must prove that

$$\theta \mathbf{x} + (1-\theta)\mathbf{y} + \eta \mathscr{D} \subset C$$

for some $\eta > 0$. Now, since $\mathbf{y} \in \overline{C}$, for any $\eta > 0$ we have

$$\begin{aligned}\theta \mathbf{x} + (1-\theta)\mathbf{y} + \eta \mathscr{D} &\subset \theta \mathbf{x} + (1-\theta)[C + \eta \mathscr{D}] + \eta \mathscr{D} \\ &= \theta[\mathbf{x} + \theta^{-1}(2-\theta)\eta \mathscr{D}] + (1-\theta)C.\end{aligned}$$

Since $\mathbf{x} \in \text{int}(C)$, we have $[\mathbf{x} + \theta^{-1}(2-\theta)\eta \mathscr{D}] \subset C$ for η sufficiently small and hence

$$\theta \mathbf{x} + (1-\theta)\mathbf{y} + \eta \mathscr{D} \subset \theta C + (1-\theta)C = C. \qquad \diamond$$

Given an arbitrary set $S \subset \mathbb{R}^n$, the smallest convex set C containing S is called the *convex hull* of S and it is denoted by $C = \text{Co}(S)$.

Proposition A.3. *Given a set $S \subset \mathbb{R}^n$, its convex hull $C = \text{Co}(S)$ consists of all convex combinations of elements of S.*

The following theorem, which constitutes a fundamental result in convex analysis, states that there exists an upper bound for the number of elements involved in the convex combinations generating the convex hull of a given set.

Theorem A.2 (Carathéodory Theorem). *Let S be any set of points in $\mathbb{R}^n$, and consider the convex hull $C = Co(S)$. Then $\mathbf{x} \in C$ if and only if $\mathbf{x}$ can be expressed as a convex combination of at most $n+1$ points in S.*

Corollary A.1. *Let $\{C_i\}_{i=1}^k$ be a collection of convex sets in $\mathbb{R}^n$ and let C be the convex hull of the union of elements in such collection, i.e.*

$$C = \text{Co}\left(\bigcup_{i=1}^{k} C_i\right).$$

Then any point in C can be expressed as a convex combinations of at most $s = \min\{k, n+1\}$ points, each belonging to a different set C_i.

Proof. By the Carathéodory Theorem, any point $\mathbf{x} \in C$ can be expressed as convex combination of at most $n+1$ elements in $\bigcup_{i=1}^k C_i$:

$$\mathbf{x} = \theta_1 \mathbf{y}_1 + \theta_2 \mathbf{y}_2 + \cdots + \theta_{n+1} \mathbf{y}_{n+1}.$$

Suppose now that two points, say $\mathbf{y}_1, \mathbf{y}_2$ for simplicity, belong to the same set C_1; then, since C_1 is a convex set, the term $\theta_1 \mathbf{y}_1 + \theta_2 \mathbf{y}_2$ can be expressed as $\theta_\star \mathbf{z}$, with

$$\theta_\star = \theta_1 + \theta_2, \quad \mathbf{z} = \frac{\theta_1}{\theta_\star}\mathbf{x}_1 + \frac{\theta_2}{\theta_\star}\mathbf{x}_2 \in C_1.$$

The above procedure can be easily generalized to the other groups of points, this showing that the point $\mathbf{x}$ can be expressed as a convex combination involving only elements belonging to different sets $C_i, i = 1, ..., k$. ◇

Given a convex set C, a subset $S \subset C$ is said to be an internal representation of C if $C = \text{Co}(S)$. An fundamental internal representation of a convex set C is the one given by its *extreme points*.

A *face* of a convex set C is a convex subset $C' \subset C$ such that every closed segment in C with a relative interior point in C' has both endpoints in C'. The zero-dimensional faces are called *extreme points*. In particular an element $\mathbf{x} \in C$ is an extreme point if and only if the only way to express $\mathbf{x}$ as a convex combination $\mathbf{x} = \theta\mathbf{y} + (1-\theta)\mathbf{z}$ with $\theta \in (0,1)$ and $\mathbf{y}, \mathbf{z} \in C$ is by taking $\mathbf{y} = \mathbf{z} = \mathbf{x}$.

Theorem A.3 (Representation Theorem). *Let C be a closed bounded convex set and let S be the set of its extreme points. Then*

$$C \equiv \text{Co}(S).$$

A.1.2 Tools from Measure Theory and Functional Analysis

A function $s = s(x)$ on a metric space X whose range consists of finitely many points in $(-\infty, \infty)$ is called *simple function*. The general expression of a simple function is given by

$$s(x) = \sum_{j=1}^{r} \alpha_i \chi_{A_i},$$

where $\alpha_i, i = 1, .., r$ are real constant coefficients and A_i are disjoint measurable subsets satisfying the identity

$$\bigcup_{j=1}^{r} A_i = X.$$

Theorem A.4. *Let $f : X \to \mathbb{R}$ be a measurable positive function. There exists a sequence of simple functions $\{s_h\}_{h=1}^{\infty}$ such that*

(a) $0 \le s_1 \le s_2 \le \cdots \le f$

(b) $\lim_{h\to\infty} s_h(x) = f(x)$ for every $x \in X$.

Proof. For $h = 1, 2, ...,$ and for $1 \le i \le h2^h$, define the sets

$$E_{h,i} = f^{-1}\left(\left[\frac{i-1}{2^h}, \frac{i}{2^h}\right)\right) \quad \text{and} \quad F_h = f^{-1}([h, \infty))$$

and put

$$s_n = \sum_{i=1}^{h2^h} \frac{i-1}{2^h} \chi_{E_{h,i}} + n\chi_{F_h}. \tag{A.1}$$

It is easy to verify that the functions s_h verify condition (a). On the other hand, for any fixed $x \in X$, we have

$$s_h(x) \geq f(x) - 2^{-h},$$

provided that h is large enough; this proves (b). ♢

Corollary A.2. *Let $\Omega \subset \mathbb{R}^n$ be a bounded closed set and $f : \Omega \to \mathbb{R}$ be a continuous function; then the simple functions s_h defined in (A.1) converge uniformly to f as h tends to infinity. In particular one has*

$$\lim_{h\to\infty} \int_\Omega |f(x) - s_h(x)| = 0.$$

The notion of simple function can be extended to vector-valued functions in a natural way; a vector-valued simple function is a function $S : X \to \mathbb{R}^m$ assuming a finite number of values and its general expression is given below

$$S(x) = \sum_{j=1}^{r} \mathbf{a}_j \chi_{A_j},$$

where $\bigcup_{j=1}^r A_j = X$, $A_i \cap A_j = \emptyset$ for $i \neq j$, and $\mathbf{a}_j \in \mathbb{R}^m$.

To conclude this overview of classical results from analysis, one may recall the general version of the implicit function theorem.

Let $\Omega \subset \mathbb{R}^{n+m}$ be an open set and let $F = F(x_1,...,x_n,y_1,...,y_m) : \Omega \to \mathbb{R}^m$ be a differentiable application. We define the Jacobian matrices

$$\frac{\partial F}{\partial x} = \begin{bmatrix} \frac{\partial F_1}{\partial x_1} & \frac{\partial F_1}{\partial x_2} & \cdots & \frac{\partial F_1}{\partial x_n} \\ \frac{\partial F_2}{\partial x_1} & \frac{\partial F_2}{\partial x_2} & \cdots & \frac{\partial F_2}{\partial x_n} \\ \vdots & \vdots & \ddots & \vdots \\ \frac{\partial F_m}{\partial x_1} & \frac{\partial F_m}{\partial x_2} & \cdots & \frac{\partial F_m}{\partial x_n} \end{bmatrix}, \quad \frac{\partial F}{\partial y} = \begin{bmatrix} \frac{\partial F_1}{\partial y_1} & \frac{\partial F_1}{\partial y_2} & \cdots & \frac{\partial F_1}{\partial y_m} \\ \frac{\partial F_2}{\partial y_1} & \frac{\partial F_2}{\partial y_2} & \cdots & \frac{\partial F_2}{\partial y_m} \\ \vdots & \vdots & \ddots & \vdots \\ \frac{\partial F_m}{\partial y_1} & \frac{\partial F_m}{\partial y_2} & \cdots & \frac{\partial F_m}{\partial y_m} \end{bmatrix}.$$

Theorem A.5. *If $(x_0, y_0) \in \Omega$ is such that*

$$F(x_0, y_0) = 0, \quad \det \frac{\partial F(x_0, y_0)}{\partial y} \neq 0$$

then one can determine a neighboorod U of x_0 in $\mathbb{R}^n$ and a neighborood V of y_0 in $\mathbb{R}^m$ such that, for any $x \in U$ there exists a unique $y = f(x) \in V$ with $F(x,y) = 0$; moreover the function $f : U \to \mathbb{R}^m$ is differentiable and verifies

$$\frac{\partial f(x)}{\partial x} = \left(\frac{\partial F(x, f(x))}{\partial y}\right)^{-1} \cdot \frac{\partial F(x, f(x))}{\partial x}.$$

A.2 Support Material for Chapter 4

This section contains some results needed in the proof of the main Theorem of Chapter 4.

Proposition A.4. *For any choice of* $d_{i,j} \lesseqgtr 0$, $i = 1,m$, $j = 1,\dots,n-m$, *constraining the plant (5.6) onto the surface* $\mathbf{s}(\mathbf{x}(k),\mathbf{x}(0),k) = 0$ *implies plant asymptotical stabilization.*

Proof. The statement can be proved making use of a standard results (reported e.g. in [80], [76]). With reference to plant (5.6), partition the state variable $\mathbf{x} = [\mathbf{x}_1^T \ x_2^T]^T$ according to the partitions of the matrices **A**,**C**,**D**. On the surface $\mathbf{s}(\mathbf{x}(k),\mathbf{x}(0),k) = 0$ given by (5.8) one has

$$x_2 = \mathbf{C}_2^{-1}\left[-\mathbf{C}_1(t)\mathbf{x}_1 + (\mathbf{C}_1(t)\mathbf{x}_1(0) + \mathbf{C}_2 x_2(0))e^{(-\bar{\lambda}t)}\right]$$

where $\mathbf{C}_1(t) \stackrel{\text{def}}{=} (\mathbf{C}_1 + \mathbf{D}_1 e^{(-\bar{\lambda}t)})$. The dynamics of the plant restricted on the sliding surface are:

$$\begin{aligned}\dot{\mathbf{x}}_1 &= \Lambda_1\mathbf{x}_1 + \Lambda_2 e^{(-\bar{\lambda}t)}\mathbf{x}_1 + \mathbf{G}_1 e^{(-\bar{\lambda}t)} + \mathbf{G}_2 e^{(-2\bar{\lambda}t)} \\ &= \Lambda(t)\mathbf{x}_1 + g(t)\end{aligned} \tag{A.2}$$

where $\Lambda_1 \stackrel{\text{def}}{=} \mathbf{A}_{1,1} - \mathbf{A}_{1,2}\mathbf{C}_2^{-1}\mathbf{C}_1$ has stable assigned eigenvalues by a suitable selection of the matrix **C**, $\Lambda_2 \stackrel{\text{def}}{=} -\mathbf{A}_{1,2}\mathbf{C}_2^{-1}\mathbf{D}_1$, $\mathbf{G}_1 \stackrel{\text{def}}{=} \mathbf{A}_{1,2}(\mathbf{C}_2^{-1}\mathbf{C}_1\mathbf{x}_1(0) + x_2(0))$, $\mathbf{G}_2 \stackrel{\text{def}}{=} \mathbf{A}_{1,2}\mathbf{C}_2^{-1}\mathbf{D}_1\mathbf{x}_1(0)$. It can be easily proved (see e.g. [80], [76], [81]) that the linear homogeneous system $\dot{\mathbf{x}}_1 = \Lambda(t)\mathbf{x}_1$ is globally exponentially stable. It follows that the state transition matrix $\Phi(t,\tau)$ of (A.2) [80], for any initial time $\tau \in [0,\infty)$ and for any initial condition $\xi \in \mathbb{R}^m$, is such that $||\Phi(t,\tau)\xi|| \le K_1 e^{-\gamma_1(t-\tau)}$ $\forall\xi$, $\forall t \ge \tau$ for suitable $\gamma_1, K_1 > 0$. Recalling [80] that the unique solution $\varphi(t,\tau,\xi)$ of (A.2) satisfying $\varphi(\tau,\tau,\xi) = \xi$ is given by $\varphi(t,\tau,\xi) = \Phi(t,\tau)\xi + \int_\tau^t \Phi(t,\eta)(\mathbf{G}_1 e^{(-\bar{\lambda}\eta)} + \mathbf{G}_2 e^{(-2\bar{\lambda}\eta)})d\eta$ one has $||\varphi(t,\tau,\xi)|| \le ||\Phi(t,\tau)\xi|| + (||\mathbf{G}_1|| + ||\mathbf{G}_2||)\int_\tau^t ||\Phi(t,\eta)||d\eta \le K_2 e^{-\gamma_2(t-\tau)}$ for suitable $K_2, \gamma_2 > 0$. This shows that the plant (5.6) restricted onto (5.8) is globally exponentially stable. ♢

Lemma A.1. *A positive constant* $\bar{\lambda}$ *can always be found such that (5.15) and (4.66) are simultaneously satisfied.*

Proof. According to the proof of the Theorem 5.2, the fulfillment of (4.66) will be imposed as follows: *i*) for $j > 1$ even, it will be imposed simultaneously $v_j \le -\bar{\lambda}$ and $\mu_j > 0$; *ii*) for $j > 1$ odd, the condition (4.66) will be imposed explicitly. A suitable constant $\bar{\lambda} > 0$ can be always found according to the procedure given below.

- It is straightforward to prove that the condition $\mu_j > 0$ for $j > 1$ is fulfilled if:

$$\frac{1}{m_r} - \frac{n}{\bar{\lambda}}\mu_{r-1} > 0 \quad \Rightarrow \quad \bar{\lambda} = n\gamma_j m_j \mu_{j-1}, \quad \gamma_j > 1 \tag{A.3}$$

- For $j = 2$, consider to impose the condition $\nu_2 \leq -\bar{\lambda}$ and $\mu_2 > 0$. The former inequality is satisfied for $0 \leq \bar{\lambda} \leq \bar{\lambda}_2$, with $\bar{\lambda}_2$ defined in (5.23). The latter corresponds to (A.3) for $j = 2$, and requires that $\bar{\lambda} > \bar{\lambda}_1$ as defined in (5.23), i.e. $\frac{1}{m_2} = \gamma_2 \frac{n}{\bar{\lambda} m_1}$; $\Rightarrow \mu_2 = (\gamma_2 - 1)\frac{n}{\bar{\lambda} m_1}$ for a suitable $\gamma_2 > 1$.
- For $j = 3$, the condition $M_1\mu_3 - \nu_3 > 0$ gives $\frac{M_1 n^2(\gamma_2-1)}{\bar{\lambda} m_1} - |a_{i,2}| - \theta_2|a_{i,1}| + \frac{\theta_1|a_{i,1}|n}{\bar{\lambda}} < \frac{M_1\bar{\lambda}}{m_3}$, i.e. $\bar{\lambda}^2 > \frac{m_3 n}{M_1}\left(\frac{M_1 n\gamma_2}{m_1} + \theta_1|a_{i,1}|\right) \stackrel{\text{def}}{=} \bar{\lambda}_3^2 > 0$.
- For $j = 4$, the condition $\nu_4 < -\bar{\lambda}$, considering that $\nu_2 < -\bar{\lambda}$, gives $|a_{i,3}| + |a_{i,4}| + \frac{\nu_2 n^2}{\bar{\lambda}^2} = |a_{i,3}| + |a_{i,4}| - \frac{n^2}{\bar{\lambda}} < -\bar{\lambda} \Rightarrow \bar{\lambda} < \bar{\lambda}_4$, while $\mu_4 > 0$ produces $\bar{\lambda} > n\mu_3 m_4$, which is satisfied if $\bar{\lambda} > n\dfrac{m_4}{m_3}$, i.e. $\dfrac{1}{m_4} > \dfrac{1}{m_3}\dfrac{n}{\bar{\lambda}}$, i.e. $\dfrac{1}{m_4} = \gamma_4\dfrac{1}{m_3}\dfrac{n}{\bar{\lambda}}$, $\gamma_4 > 1$.
- For $j = 5$, the condition $M_1\mu_5 - \nu_5 > 0$, proceeding similarly to $i = 3$, is imposed if one requires that $\frac{n^2 M_1(\gamma_4-1)}{\bar{\lambda}^2 m_3} - \frac{\nu_4}{\bar{\lambda}} < \frac{M_1}{m_5}$. Recalling that $\nu_4 = |a_{i,3}| + |a_{i,4}| + \dfrac{n^2\nu_2}{\bar{\lambda}^2}$, a stronger condition than the previous one is $\frac{n^2 M_1(\gamma_4-1)}{\bar{\lambda}^2 m_3} + \frac{\theta_1|a_{i,1}|n}{\bar{\lambda}^4} < \frac{M_1}{m_5}$ and, using the inequality $\theta_1|a_{1,1}|n < \bar{\lambda}^2\frac{M_1}{m_3}$ coming from the step $j = 3$, one has $\bar{\lambda}^2 > \frac{m_5}{m_3}(\gamma_4 n^2 + 1) \stackrel{\text{def}}{=} \bar{\lambda}_5^2 > 0$. All further odd j's can be treated similarly.
- For $j = 6$, the conditions $\mu_6 > 0$ and $\nu_6 < 0$, imposed similarly to $j = 4$, give $\bar{\lambda} > n\frac{m_6}{m_5}$ and $\bar{\lambda} < \bar{\lambda}_6$, for a suitable $\bar{\lambda}_6 > 0$. All further even j's can be treated similarly.

Summing up, $\bar{\lambda}$ has to satisfy:

$$\bar{\lambda} < \min\{\bar{\lambda}_2, \bar{\lambda}_4, \ldots,\} \stackrel{\text{def}}{=} \bar{\lambda}_M;$$
$$\bar{\lambda} > \max\{\bar{\lambda}_1, \bar{\lambda}_3, \bar{\lambda}_5, \ldots,\} \stackrel{\text{def}}{=} \bar{\lambda}_m \tag{A.4}$$
$$\frac{1}{m_2} - \gamma_2\frac{n}{\bar{\lambda} m_1}; \frac{1}{m_4} = \gamma_4\frac{n}{\bar{\lambda} m_3} \ldots; \quad \gamma = \max\{\gamma_2, \gamma_4, \ldots\} > 1 \tag{A.5}$$

The value $\bar{\lambda}_M$ in (A.4), can be computed exactly, as depends only on plant parameters. To ensure that an admissible interval of values for $\bar{\lambda}$ exists, one has to impose that $\bar{\lambda}_1 < \bar{\lambda}_M; \bar{\lambda}_3 < \bar{\lambda}_M; \bar{\lambda}_5 < \bar{\lambda}_M, \ldots$. Since $\bar{\lambda}_1, \bar{\lambda}_3$ depend inversely on m_1, $\bar{\lambda}_5$ depends inversely on m_3 and so on, the above conditions produce a lower bound on odd coefficients m_j, i.e. $m_1 > m_{1,m}; \quad m_3 > m_{3,m}; \ldots m_j > m_{j,m}; \quad j\ odd,\ j = 1, \ldots, n-m$ Note that for the largest index $k = n - m + 1$, the only requirement for the corresponding m_{n-m+1} is simply (5.15). It remains now to show that the constraint (5.15) is satisfied. To this purpose, choosing $m_j > n$, $j = 1, \ldots, n-m+1$, then (5.15) is fulfilled (the converse is not true). Fix m_j, for j odd and $j \leq n - m$ according to

$$m_1 > \max\{n, m_{1,m}\}; \quad m_3 > \max\{n, m_{3,m}\}; \ldots;$$
$$m_j > \max\{n, m_{j,m}\}; \quad j\ odd,\ j \leq n - m \tag{A.6}$$

Recalling (A.5), the constraint (5.15) gives $n\frac{\gamma}{\bar{\lambda}}\left(\frac{1}{m_1}+\frac{1}{m_3}+\dots\right)\leq 1-\frac{1}{m_1}-\frac{1}{m_3}-\dots$ i.e. $\bar{\lambda}\geq\frac{\gamma\left(\frac{1}{m_1}+\frac{1}{m_3}+\dots\right)n}{1-\frac{1}{m_1}-\frac{1}{m_3}-\dots}$. To ensure that $\bar{\lambda}$ belongs to the feasibility interval $\bar{\lambda}>\bar{\lambda}_m$, it is enough to impose

$$\frac{1}{m_1}+\frac{1}{m_3}+\dots+\frac{1}{m_k}>\frac{\bar{\lambda}_m}{n\gamma}\left\{1-\frac{1}{m_1}-\frac{1}{m_3}-\dots-\frac{1}{m_k}\right\}$$

i.e. $n\gamma\left(\frac{1}{m_1}+\frac{1}{m_3}+\dots\right)>\bar{\lambda}_m\left\{1-\frac{1}{m_1}-\frac{1}{m_3}\dots\right\}$. It is enough to set:

$$\gamma\geq\max\left\{1,\frac{\bar{\lambda}_m\left(1-\frac{1}{m_1}-\frac{1}{m_3}-\dots\right)}{n\left(\frac{1}{m_1}+\frac{1}{m_3}+\dots\right)}\right\} \tag{A.7}$$

◇

References

1. Hu, T., Lin, Z., Pitsillides, A.: In: Kapila, V., Grigoriadis, K. (eds.) Actuator Saturation Control, (Marcel Dekker)
2. Hu, T., Lin, Z.: Control Systems with Actuator Saturation. Kirkhauser (2001)
3. Sontag, E.: Int. J. Control 39, 181 (1984)
4. Fuller, A.: Int. J. Control 10(4), 457 (1969)
5. Sontag, E., Sussmann, H.: In: Proc. IEEE Conf. Decision and Cont., pp. 3414–3416 (1990)
6. Sussmann, H., Yang, Y.: In: Proc. IEEE Conf. Decision and Cont., pp. 70–72 (1991)
7. Lin, Z.: Automatica 34, 897 (1998)
8. Teel, A.: Sys. Contr. Lett. 18, 165 (1992)
9. Sontag, E., Sussman, H., Yang, Y.: IEEE Trans. Autom. Contr. 39, 2411 (1994)
10. Lin, Z., Saberi, A.: Sys. Contr. Lett. 21, 225 (1993)
11. Lin, Z., Saberi, A.: Sys. Contr. Lett. 24, 125 (1995)
12. Lin, Z., Saberi, A., Teel, A.: In: Proc. IEEE Conf. Decision and Cont, pp. 285–289 (1995)
13. Gokcek, C., Kabamba, P., Merkoov, S.: IEEE Trans. Autom. Contr. AC-46(10), 1529 (2001)
14. Bernstein, D., Michel, A.: Int. J. Rob. Nonlin. Cont. 5(5), 375 (1995)
15. Hu, T., Lin, Z., Qiu, L.: IEEE Trans. Autom. Contr. AC-46(6), 973 (2001)
16. Hippe, P., Wurmthaler, C.: Automatica 35, 689 (1999)
17. Grimm, G., Teel, A., Zaccarian, L.: Automatica 40, 1987 (2004)
18. Teel, A.: Int. J. Rob. Nonlin. Cont. 9, 701 (1999)
19. Chen, B., Lee, T., Venkataramanan, V.: IEEE Trans. Autom. Contr. AC-48(3), 427 (2003)
20. Lin, Z.: Sys. Contr. Lett. 29, 215 (1997)
21. Paim, C., Tarbouriech, S., da Silva Jr., J.M.G., Castelan, E.B.: In: Proc. IEEE Conf. Decision and Cont., pp. 4148–4153 (2002)
22. Saberi, A., Lin, Z., Teel, A.: IEEE Trans. Autom. Contr. AC-41(3), 368 (1996)
23. Fang, H., Lin, Z.: IEEE Trans. Autom. Contr. AC-51(7), 1177 (2006)
24. Hu, T., Lin, Z., Qiu, L.: Sys. Cont. Lett. 47, 65 (2002)
25. Wang, X., Saberi, A., Stoorvogel, A., Roy, S., Sannuti, P.: Int. J. Control 82, 1870 (2009)
26. Coutinho, D., Gomes da Silva, J.J.M.: IET J. Control and Appl. 4, 315 (2010)
27. González, A., Odloak, D.: Automatica 45, 1080 (2009)
28. Barmish, B.R., Schmitendorf, W.E.: SIAM J. Control and Opt. 18, 327 (1980)
29. Saberi, A., Hou, P., Stoorvogel, A.A.: IEEE Trans. Autom. Contr. 45(6), 1042 (2000)
30. Stoorvogel, A.A., Saberi, A., Shi, G.: Automatica 40, 1481 (2004)
31. LeMay, J.L.: IEEE Trans. Autom. Cont. 9, 346 (1964)
32. Corradini, M., Cristofaro, A., Giannoni, F.: IET Cont. Theory Appl. 5(5), 744 (2011)
33. Rockafellar, R.T.: Convex analysis. University Press (1997)
34. Rudin, W.: Real and complex analysis. Mc-Graw Hill (1998)
35. Hu, T., Miller, D., Qiu, L.: Automatica 38, 2009 (2002)
36. Corradini, M., Cristofaro, A., Giannoni, F.: In: European Cont. Conf. (2009)
37. Blanchini, F.: Automatica 35, 1747 (1999)
38. Dorea, C., Hennet, J., Optim, J.: Theory and Appl. 103, 521 (1999)
39. Bartolini, G., Pisano, A., Usai, E.: IEEE Trans. Autom. Cont. 46, 1826 (2001)
40. Corradini, M.L., Orlando, G.: Automatica 43(1), 88 (2007)
41. Corradini, M., Cristofaro, A., Orlando, G.: IEEE Trans. Autom. Cont. 419, 419 (2010)
42. Yilmaz, C., Hurmuzlu, Y.: ASME J. Dyn. Sys. Meas. and Cont. 122(4), 753 (2000)
43. Utkin, V., Shi, J.: In: Proc. IEEE Conf. IEEE Conf. Decision Cont., pp. 4591–4596 (1996)

44. Tarbouriech, S., da Silva Jr., J.G.: IEEE Trans. Autom. Contr. 45, 105 (2000)
45. Cao, Y., Lin, Z., Hu, T.: IEEE Trans. Circ. Sys. I 49, 233 (2002)
46. Benzaouia, A.: IEEE Trans. Autom. Cont. 39, 2091 (1994)
47. Castelan, E., da Silva Jr., J.G., Cury, J.: IEEE Trans. Autom. Cont. 38, 249 (1993)
48. Cristea, M.: Journal of Ineq. Pure and Appl. Mathematics, 128–143 (2007)
49. Corradini, M., Cristofaro, A., Giannoni, F.: Far. East J. of Math. Science 48, 427 (2010)
50. Utkin, V.: Sliding modes in control optimization. Springer, Berlin (1992)
51. Young, K., Utkin, V., Ozguner, U.: IEEE Trans. Cont. Sys. Technol. 7(3), 328 (1999)
52. Corradini, M.L., Orlando, G.: Automatica 43, 88 (2007)
53. Corradini, M., Cristofaro, A., Orlando, G.: In: Proc. IEEE Conf. Decision Cont. (2008)
54. Antsaklis, P., Michel, A.: Linear Systems. Birkhäuser, Boston (2006)
55. Rugh, W.: Linear System Theory. Prentice-Hall (1993)
56. Emelyanov, S., Korovin, S., Mamedov, I.: Variable Structure Control Systems: Discrete and Digital. Mir Publishers, Moscow (1995)
57. Edwards, C., Spurgeon, S.K.: Sliding Mode Control: Theory and Applications. Taylor & Francis, Abington (1998)
58. Slotine, J., Sastry, S.: Int. J. Control 38(2), 465 (1983)
59. G.W.W.Y.H.A.: IEEE Trans. on Ind. Electron. 42(2), 117 (1995)
60. Gomez Da Silva, J., Tarbouriech, S.: IEEE Trans. Autom. Contr. AC-46, 119 (2001)
61. Pittet, C., Tambouriech, S., Burgat, C.: In: Proc. IEEE Conf. Decision and Cont., pp. 4518–4523 (1997)
62. Hindi, H., Boyd, S.: In: Proc. IEEE Conf. Decision and Cont., pp. 903–908 (1998)
63. Hu, T., Lin, Z., Chen, B.: Sys. Contr. Lett. 45, 97 (2002)
64. Cao, Y., Lin, Z.: Automatica 39, 1235 (2003)
65. Teel, A.: IEEE Trans. Autom. Contr. AC-40, 96 (1995)
66. Choi, J.: Sys. Contr. Lett. 36, 241 (1999)
67. Hu, T., Lin, Z.: In: Proc. Amer. Cont. Conf. (2000)
68. Choi, J.: In: Proc. Am. Cont. Conf., pp. 4926–4929 (2001)
69. Saberi, A., Han, J., Stoorvogel, A.A.: Automatica 38, 639 (2002)
70. Hou, P., Saberi, A., Lin, Z.: In: Proc. IEEE Conf. Decis. Cont., vol. 5 (1997)
71. Stoorvogel, A.A., Saberi, A., Shi, G.: In: Proc. IEEE Conf. Decision and Cont., vol. 3 (1999)
72. Wang, Y., Cao, Y., Li, S., Sun, Y.: Proc. Am. Cont. Conf. (2006)
73. Cao, Y., Lin, Z., Ward, D.: IEEE Trans. Autom. Cont. 47(1), 140 (2002)
74. Tarbouriech, S., Garcia, G.: Control of uncertain systems with bounded inputs. Springer-Verlag New York, Inc., Secaucus (1997)
75. Zuo, Z., Jia, Z., Wang, Y., Zhao, H., Zhang, G.: In: Proc. American Cont. Conf., pp. 3632–3637 (2008)
76. Michel, A., Hou, L., Liu, D.: Stability of dynamical systems. Birkhäuser (2008)
77. Utkin, V.: IEEE Trans. on Autom. Cont. 22(2), 212 (1977)
78. Corradini, M., Orlando, G.: IEEE Trans. Autom. Contr. AC-43(9), 1329 (1998)
79. Kolmogorov, A., Fomine, S.: Eléments de la théorie des fonctions et de l'analyse fonctionbelle, Ellipses (MIR), Paris, Moscow (1994)
80. Miller, R., Michel, A.: Ordinary Differential equations. Dover Publications, Inc. (1982)
81. Slotine, J., Li, W.: Applied nonlinear control. Prentice Hall International Inc., Upper Saddle River (1991)

Index

Lecture Notes in Control and Information Sciences

Edited by M. Thoma, F. Allgöwer, M. Morari

Further volumes of this series can be found on our homepage:
springer.com

Vol. 424: Corradini, M.L.; Cristofaro, A.; Giannoni, F.; Orlando, G.:
Control Systems with Saturating Inputs
134 p. 2012 [978-1-4471-2505-1]

Vol. 423: Sipahi, R.; Vyhlídal, T.; Niculescu, S.-I.; Pepe, P. (Eds.)
Time Delay Systems: Methods, Applications and New Trends
442 p. 2012 [978-3-642-25220-4]

Vol. 422: Kozłowski, K.R. (Ed.)
Robot Motion and Control 2011
418 p. 2012 [978-1-4471-2342-2]

Vol. 420: Trinh, H.; Fernando, T.:
Functional Observers for Dynamical Systems
218 p. 2012 [978-3-642-24063-8]

Vol. 419: Samy, I.; Gu, D.-W.:
Fault Detection and Flight Data Measurement
170 p. 2012 [978-3-642-24051-5]

Vol. 418: Alberer, D.; Hjalmarsson, H.; del Re, L.:
Identification for Automotive Systems
348 p. 2012 [978-1-4471-2220-3]

Vol. 417: Johansson, R.; Rantzer A.:
Distributed Decision Making and Control
412 p. 2012 [978-1-4471-2264-7]

Vol. 416: Varga, A.; Hansson, A.; Puyou, G.:
Optimization Based Clearance of Flight Control Laws
451 p. 2012 [978-3-642-22626-7]

Vol. 415: Chesi, G.:
Domain of Attraction
283 p. 2011 [978-0-85729-958-1]

Vol. 414: Imine, H.; Fridman, L.; Shraim, H.; Djemai, M.:
Sliding Mode Based Analysis and Identification of Vehicle Dynamics
127 p. 2011 [978-3-642-22223-8]

Vol. 413: Eleftheriou, E.; Reza Moheimani, S.O.:
Control Technologies for Emerging Micro and Nanoscale Systems
289 p. 2011 [978-3-642-22172-9]

Vol. 412: Fridman, L.; Moreno, J.; Iriarte, R.:
Sliding Modes after the first Decade of the 21st Century
XXX p. 2011 [978-3-642-22163-7]

Vol. 411: Kaczorek, T.;
Selected Problems of Fractional Systems Theory
344 p. 2011 [978-3-642-20501-9]

Vol. 410: Bourlès, H., Marinescu, B.:
Linear Time-Varying Systems
637 p. 2011 [978-3-642-19726-0]

Vol. 409: Xia, Y., Fu, M., Liu, G.-P.:
Analysis and Synthesis of Networked Control Systems
198 p. 2011 [978-3-642-17924-2]

Vol. 408: Richter, J.H.;
Reconfigurable Control of Nonlinear Dynamical Systems
291 p. 2011 [978-3-642-17627-2]

Vol. 407: Lévine, J., Müllhaupt, P.:
Advances in the Theory of Control, Signals and Systems with Physical Modeling
380 p. 2010 [978-3-642-16134-6]

Vol. 406: Bemporad, A., Heemels, M., Johansson, M.:
Networked Control Systems
appro. 371 p. 2010 [978-0-85729-032-8]

Vol. 405: Stefanovic, M., Safonov, M.G.:
Safe Adaptive Control
appro. 153 p. 2010 [978-1-84996-452-4]

Vol. 404: Giri, F.; Bai, E.-W. (Eds.):
Block-oriented Nonlinear System Identification
425 p. 2010 [978-1-84996-512-5]

Vol. 403: Tóth, R.;
Modeling and Identification of Linear Parameter-Varying Systems
319 p. 2010 [978-3-642-13811-9]

Vol. 402: del Re, L.; Allgöwer, F.; Glielmo, L.; Guardiola, C.; Kolmanovsky, I. (Eds.):
Automotive Model Predictive Control
284 p. 2010 [978-1-84996-070-0]

Vol. 401: Chesi, G.; Hashimoto, K. (Eds.):
Visual Servoing via Advanced Numerical Methods
393 p. 2010 [978-1-84996-088-5]

Vol. 400: Tomás-Rodríguez, M.; Banks, S.P.:
Linear, Time-varying Approximations to Nonlinear Dynamical Systems
298 p. 2010 [978-1-84996-100-4]

Vol. 399: Edwards, C.; Lombaerts, T.; Smaili, H. (Eds.):
Fault Tolerant Flight Control
appro. 350 p. 2010 [978-3-642-11689-6]

Vol. 398: Hara, S.; Ohta, Y.; Willems, J.C.; Hisaya, F. (Eds.):
Perspectives in Mathematical System Theory, Control, and Signal Processing
appro. 370 p. 2010 [978-3-540-93917-7]

Vol. 397: Yang, H.; Jiang, B.; Cocquempot, V.:
Fault Tolerant Control Design for Hybrid Systems
191 p. 2010 [978-3-642-10680-4]

Vol. 396: Kozlowski, K. (Ed.):
Robot Motion and Control 2009
475 p. 2009 [978-1-84882-984-8]

Vol. 395: Talebi, H.A.; Abdollahi, F.; Patel, R.V.; Khorasani, K.:
Neural Network-Based State Estimation of Nonlinear Systems
appro. 175 p. 2010 [978-1-4419-1437-8]

Vol. 394: Pipeleers, G.; Demeulenaere, B.; Swevers, J.:
Optimal Linear Controller Design for Periodic Inputs
177 p. 2009 [978-1-84882-974-9]

Vol. 393: Ghosh, B.K.; Martin, C.F.; Zhou, Y.:
Emergent Problems in Nonlinear Systems and Control
285 p. 2009 [978-3-642-03626-2]

Vol. 392: Bandyopadhyay, B.; Deepak, F.; Kim, K.-S.:
Sliding Mode Control Using Novel Sliding Surfaces
137 p. 2009 [978-3-642-03447-3]

Vol. 391: Khaki-Sedigh, A.; Moaveni, B.:
Control Configuration Selection for Multivariable Plants
232 p. 2009 [978-3-642-03192-2]

Vol. 390: Chesi, G.; Garulli, A.; Tesi, A.; Vicino, A.:
Homogeneous Polynomial Forms for Robustness Analysis of Uncertain Systems
197 p. 2009 [978-1-84882-780-6]

Vol. 389: Bru, R.; Romero-Vivó, S. (Eds.):
Positive Systems
398 p. 2009 [978-3-642-02893-9]

Vol. 388: Jacques Loiseau, J.; Michiels, W.; Niculescu, S-I.; Sipahi, R. (Eds.):
Topics in Time Delay Systems
418 p. 2009 [978-3-642-02896-0]

Vol. 387: Xia, Y.; Fu, M.; Shi, P.:
Analysis and Synthesis of Dynamical Systems with Time-Delays
283 p. 2009 [978-3-642-02695-9]

Vol. 386: Huang, D.; Nguang, S.K.:
Robust Control for Uncertain Networked Control Systems with Random Delays
159 p. 2009 [978-1-84882-677-9]

Vol. 385: Jungers, R.:
The Joint Spectral Radius
144 p. 2009 [978-3-540-95979-3]

Vol. 384: Magni, L.; Raimondo, D.M.; Allgöwer, F. (Eds.):
Nonlinear Model Predictive Control
572 p. 2009 [978-3-642-01093-4]

Vol. 383: Sobhani-Tehrani E.; Khorasani K.;
Fault Diagnosis of Nonlinear Systems Using a Hybrid Approach
360 p. 2009 [978-0-387-92906-4]

Vol. 382: Bartoszewicz A.; Nowacka-Leverton A.;
Time-Varying Sliding Modes for Second and Third Order Systems
192 p. 2009 [978-3-540-92216-2]

Vol. 381: Hirsch M.J.; Commander C.W.; Pardalos P.M.; Murphey R. (Eds.)
Optimization and Cooperative Control Strategies: Proceedings of the 8th International Conference on Cooperative Control and Optimization
459 p. 2009 [978-3-540-88062-2]

Vol. 380: Basin M.
New Trends in Optimal Filtering and Control for Polynomial and Time-Delay Systems
206 p. 2008 [978-3-540-70802-5]

Vol. 379: Mellodge P.; Kachroo P.;
Model Abstraction in Dynamical Systems: Application to Mobile Robot Control
116 p. 2008 [978-3-540-70792-9]

Vol. 378: Femat R.; Solis-Perales G.;
Robust Synchronization of Chaotic Systems Via Feedback
199 p. 2008 [978-3-540-69306-2]

Vol. 377: Patan K.
Artificial Neural Networks for the Modelling and Fault Diagnosis of Technical Processes
206 p. 2008 [978-3-540-79871-2]